SLOPE STABILITY AND EROSION CONTROL:
ECOTECHNOLOGICAL SOLUTIONS

Slope Stability and Erosion Control: Ecotechnological Solutions

Edited by

Joanne E. Norris
Halcrow Group Ltd.,
Peterborough, U.K.

Alexia Stokes
INRA,
Montpellier, France

Slobodan B. Mickovski
Jacobs UK Ltd.,
Glasgow, U.K.

Erik Cammeraat
IBED, University of Amsterdam,
The Netherlands

Rens van Beek
Utrecht Centre of Geosciences,
Utrecht University, The Netherlands

Bruce C. Nicoll
Forest Research, Roslin, U.K.

and

Alexis Achim
Faculté de Foresterie et de Géomatique,
Université Laval, Québec, Canada

Springer

Library of Congress Control Number: 2008921372

ISBN 978-1-4020-6675-7 (HB)
ISBN 978-1-4020-6676-4 (e-book)

Published by Springer,
P.O. Box 17, 3300 AA Dordrecht, The Netherlands.

www.springer.com

Cover Legend

1. Slope failure as a result of windthrow in a mixed stand of *Tsuga heterophylla, Thuja plicata* and *Abies amabilis* at Northern Vancouver Island, British Columbia, Canada. *Reproduced by kind permission of Steve Mitchell, Faculty of Forestry University of British Columbia Vancouver, BC, Canada.*
2. The influences of vegetation on a slope. *Reproduced by kind permission of Joanne E. Norris, Halcrow Group Ltd., Endeavour House, Forder Way, Cygnet Park, Hampton, Peterborough, U.K.*
3. Live staking using cuttings. *Reproduced by kind permission of Luc Jouneau, INRA, Domaine de Vilvert, 78352 Jouy-en-Josas cedex, France.*
4. *Fagus sylvatica* root network exposed through erosive processes, on the east bank of the River Almond at Cramond near Edinburgh, UK. *Reproduced by kind permission of Bruce C. Nicoll, Forest Research, Northern Research Station, Roslin, Midlothian, UK*

Printed on acid-free paper

TABLE OF CONTENTS

Chapter 1

INTRODUCTION TO ECOTECHNOLOGICAL SOLUTIONS

Alexia Stokes[1], Joanne E. Norris[2,3], John R. Greenwood[3]

[1] INRA, AMAP, A A-51/PS2, Boulevard de la Lironde, 34398 Montpellier cedex 5, France, [2] Halcrow Group Limited, Endeavour House, Forder Way, Cygnet Park, Hampton, Peterborough, PE7 8GX, U.K., [3] School of Architecture, Design and Built Environment, Nottingham Trent University, Burton Street, Nottingham, NG1 4BU, U.K.

Abstract: *We introduce the terminology used in this book and outline the scientific principles behind the definitions given for ecotechnology, eco- and ground bio-engineering. We focus on the use of restoration and management techniques for slopes prone to shallow mass movement and erosion through natural events such as storms. The use of protection forests is discussed, along with their mechanical stability during wind storms, landslides and rockfall events. Which ecotechnological solution to use in any given situation is outlined, depending on the scale of the problem, economics and the consequences of action and inaction.*

Key words: eco-engineering, ground bio-engineering, landslides, erosion, rockfall, storms

1. INTRODUCTION

"**Ecotechnology** is the use of technological means for ecosystem management, based on deep ecological understanding, to minimize the costs of measures and their harm to the environment" (Straskraba 1993). The science of ecotechnology is similar to that called "ecological engineering," which in turn has been described as "the management of nature" (Odum 1971), or as "the proactive design of sustainable ecosystems which integrate human society with its natural environment, for the benefit of both" (Mitsch 1996; Painter 2003; Mitsch and Jørgensen 2004). Ecological engineering involves mostly creation and restoration of ecosystems whereas ecotechnology encompasses the management of ecosystems (Mitsch and Jørgensen 2004). Both subjects have largely been devoted to the sustainability of wetlands,

J.E. Norris et al. (eds.), Slope stability and erosion control: Ecotechnological solutions, 1–8.

wastewater and aquaculture (Painter 2003), but can be applied to a larger range of environments. In this book, we will focus on the restoration or protection of sites using eco- and ground bio-engineering techniques, both of which fall within the science of ecotechnology. Eco-engineering has recently been defined as the long-term, ecological strategy to manage a site with regard to natural or man-made hazards (Stokes et al. 2004). For natural slopes, such hazards can be mass movement of soil, e.g., landslides, avalanches and rockfall, or erosion, e.g., sheet and gully erosion or river bank erosion. By combining ground bio-engineering techniques with long-term solutions, slopes can be managed effectively to minimize the risk of failure.

Ground bio-engineering methods integrate civil engineering techniques with natural or man-made materials to obtain fast, effective and economic methods of protecting, restoring and maintaining the environment (Schiechtl 1980; Coppin and Richards 1990; Gray and Sotir 1996). The use of, e.g., geotextiles or brush mattressing to arrest soil run-off and the planting of fast-growing herbaceous species to fix soil, are typical ground bio-engineering techniques. The correct choice of plant material is difficult, as knowledge is required concerning the ability of the plant to grow on a particular site, and also the efficiency of the root system in fixing and reinforcing soil on an unstable slope. Although such information may be available for a particular species, its performance in the long-term also needs to be known, e.g., grasses often die back in summer and should be combined with shrubs so as to avoid slippage or erosion problems. Shade intolerant species will also decline as shrubs and trees grow taller over a longer period of time. Long-term solutions therefore need to include the use of appropriate management strategies and the employment of Decision Support Systems (DSS). Such tools could also be integrated into Geographic Information Systems (GIS) to predict future risks. Such management techniques are particularly effective in large-scale areas in Europe, e.g., ski resorts, mountain slopes and forest stands (Dorren and Seijmonsbergen 2003).

1.1 Using eco- and ground bio-engineering techniques

Examples of where eco-engineering techniques would be most useful are in situations whereby human safety is not an immediate issue, the site is large-scale, or where protecting structures are already in place, e.g., rock trap nets, avalanche barriers and gabion walls. When deciding to carry out eco-engineering techniques on an unstable slope, the engineer must first determine the nature of the slope, type of soil, type of native or desired vegetation and the likelihood of any catastrophic event occurring which would decrease slope stability during the restoration time (Figure 1.1). If the risk of danger to human life and infrastructures is low, the engineer must consider the size of the site and costs to be incurred throughout the life of the project. If the site is on a small-scale and the cost of construction, e.g., fascines, live stakes and branch nets, planting and upkeep is equal to the economic, aesthetic and safety gain at the end of the project, ground bio-engineering techniques can be considered. If the site is large-scale, e.g., a mountain slope, the expenses incurred in carrying out certain bio-engineering techniques may be too high for the gain produced, and eco-engineering techniques may be used. However, it must be remembered that any gain as a result of an eco-engineering project will only be in the long-term.

Typical eco-engineering practices may include the use of DSS (Gardiner and Quine 2000; Mickovski et al. 2005; Mickovski and van Beek 2006, see Chapter 8) to determine how and when to plant depending on soil and slope type and the hazards to which the site is exposed. Management strategies are then proposed for the upkeep of the site. For example, a mountain protection forest should consist of broadleaf species, the number of wild ungulates should be limited and thinning and felling should be carried out with care (Motta and Haudemand 2000). Similarly, in conifer forests subjected to frequent storms, the upwind border of the stand could be planted with broadleaf species and pruned to create a 'ramp', or shelterbelt type structure. Such a structure would cost little to maintain and would allow the prevailing wind to pass over the plantation, rather than penetrate into the stand (Quine et al. 1995).

Eco-engineering is beginning to emerge as a future research area in Europe which engineers and ecologists should consider both in education and application (Stokes et al. 2007). Human activity over the last 100 years has been concerned with increasing productivity through technological progress, at the cost of environmental degradation (Painter 2003). It is now necessary to repair this damage, although with limited resources, many countries are unable to invest heavily in environmental restoration of degraded lands. Eco-engineering techniques can therefore provide a low-cost, long-term solution in certain cases.

As mentioned previously, ground bio-engineering is defined as the use of living plant materials to perform some engineering function, from simple erosion control with grass and legume seeding or more complex slope stabilisation with willows (*Salix* sp.) and other plants (Schiechtl 1980). The response is fast which is particularly important for stabilizing a denuded slope.

The function of vegetation in bio-engineering can be divided into four groups (Schiechtl and Stern 1996), which are:

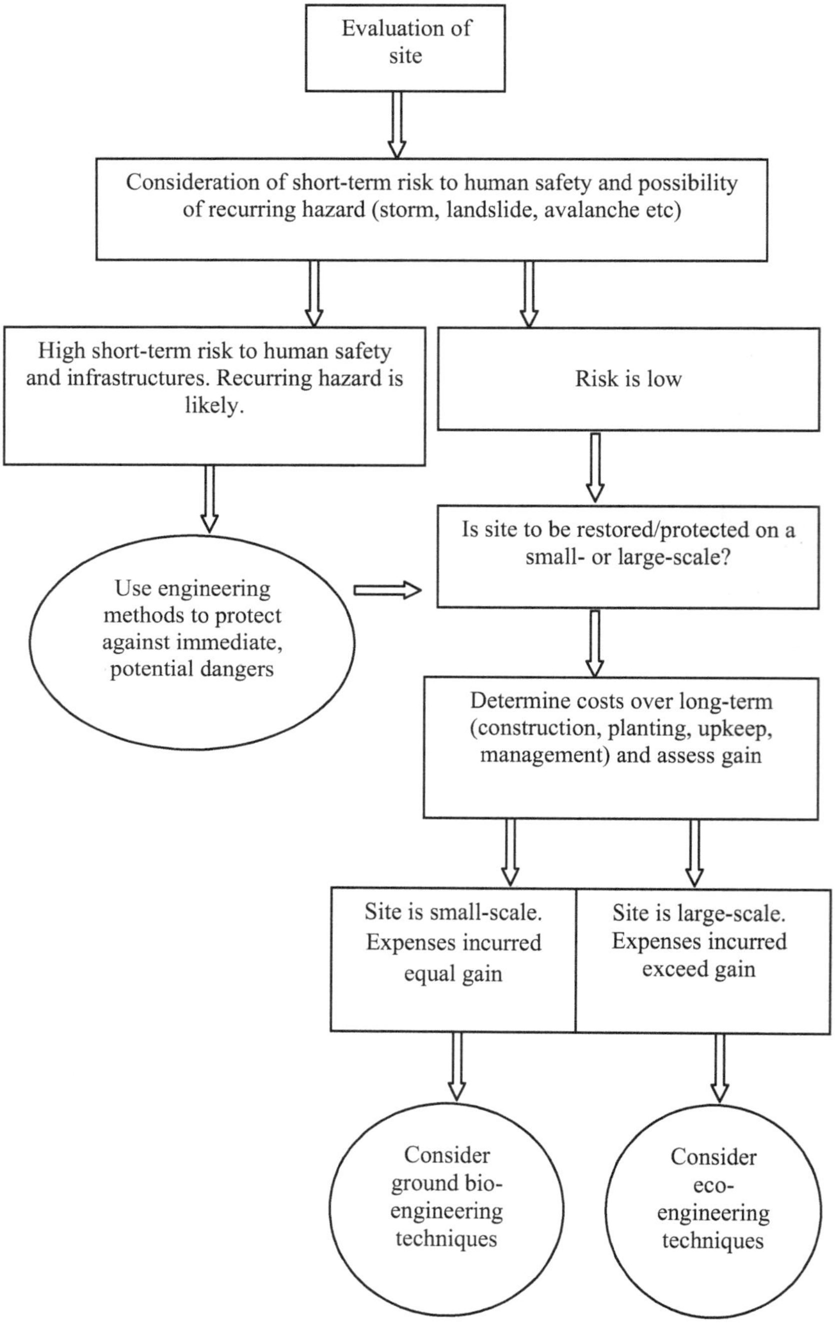

Figure 1-1. When considering the implementation of ground bio- or eco-engineering techniques, the engineer must take into account the potential dangers, size, cost and gain of the project.

1. *Soil protection techniques* rapidly protect the soil, by means of their covering action, from surface erosion and degradation. Such techniques improve water retention capacity and promote biological soil activity.
2. *Ground stabilising techniques* are designed to reduce or eliminate mechanical disturbing forces due to the soil mass. These techniques stabilise and secure slopes liable to slides by means of root penetration, decreased pore water pressure through transpiration and improved drainage. In principle, they consist of linear or single point systems of shrubs and trees.
3. *Combined construction techniques* shore up and secure unstable slopes and embankments by combining the use of live plants with inert materials (stone, concrete, wood, steel, and geosynthetics). This method increases the effectiveness and life expectancy of the measures employed.
4. *Supplementary construction techniques* comprise seeding and plantings in the widest sense of the word; they serve to secure the transition from the construction stage to the completed project.

Pioneering woody species are of particular importance in the development of ground bio-engineering systems. This group of plants represents the succession bridge between the herbaceous initial colonisers (seeded grasses and legumes) of a disturbed site and later seral types and thus plays a key role in succession advancement of the site (Polster 2003). Woody vegetation improves the hydrology and mechanical stability of slopes through root reinforcement and surface protection (Sotir 2002).

The role of vegetation in stabilising slopes is not limited to general planting techniques. One aspect of ground bio-engineering is to use living plant material to build structures to stabilise the problem site. All construction materials must be strong enough to withstand the forces acting on them. Since it is the intention to build structures of living materials, these materials must sprout and grow, therefore the materials must be in a condition that will promote their subsequent growth. Plant material is typically in the form of stem cuttings when planted and must therefore be capable of forming new roots and shoots (Polster 2002).

By using vegetation in the structure it is possible to manipulate the depth at which rooting occurs. For example, live willow stakes can be planted at a depth of 2.0 m below the surface as long as anaerobic conditions are not present (Steele et al. 2004). With traditional planting methods, roots would not normally reach this depth.

There are limitations though to ground bio-engineering methods and include:

1. Installation is often limited to the plants' dormant season, when site conditions may limit access, e.g., heavy snowfall or waterlogging.
2. The availability of locally adapted plants may be limited.
3. Labour needs are intensive and skilled, experienced labour may not be available.
4. Labourers may not be familiar with ground bio-engineering principles and designs, so upfront training may be required.
5. Alternative civil engineering practices such as soil nailing and geosynthetic reinforcement, which have well defined engineering parameters are widely used, marketed and are more commonly accepted by society and contractors (Franti 1996) especially for stabilising infrastructure slopes.

2. HOW TO USE THIS BOOK

This book has been written to provide non-specialists with the information needed to characterize an unstable slope and to decide how best to restore and/or manage the site in the long-term. Chapters 2 and 3 explain how to describe a natural or man-made slope and provide information on the different types of mass wasting which can be found. How plants reinforce soil on unstable slopes is presented in Chapter 4, with an in-depth description of root system mechanical and morphological properties. In Chapter 5, the authors discuss the principles of hazard assessment on slopes prone to mass movement and erosion. Not only is soil movement described, but tree stability during wind storms is explained, a factor which can seriously aggravate soil movement on forested slopes. Engineers require information about which species to plant on a given slope, and a comprehensive list is provided in Chapter 6. On slopes where rapid remedial measures need carrying out, ground bio-engineering methods can be used and a wide selection is presented in Chapter 7, along with the long-term management of forests against storms and rockfall. Finally, perspectives for future eco-technological research are given in Chapter 8.

3. REFERENCES

Coppin NJ, Richards IJ (1990) Use of Vegetation in Civil Engineering. CIRIA, Butterworths, London

Dorren LKA, Seijmonsbergen AC (2003) Comparison of three GIS-based models for predicting rockfall runout zones at a regional scale. Geomorphology 56:49-64

Franti TG (1996) Bioengineering for hillslope, streambank and lakeshore erosion control. http://www.ianr.unl.edu/pubs/soil/g1307.htm, 1-7

Gardiner BA, Quine CP (2000) Management of forests to reduce the risk of abiotic damage – a review with particular reference to the effects of strong winds. For Ecol Manag 135:261-277

Gray DH, Sotir RB (1996) Biotechnical and Soil Bioengineering Slope Stabilization: A Practical Guide for Erosion Control. Wiley & Sons, Inc., New York

Mickovski SB, Stokes A, van Beek LPH (2005) A decision support tool for windthrow hazard assessment and prevention. For Ecol Manag 216:64-76

Mickovski SB, van Beek LPH (2006) A decision support system for the evaluation of eco-engineering strategies for slope protection. Geotech Geol Eng 24:483-498

Mitsch WJ (1996) Ecological engineering: a new paradigm for engineers and ecologists. In: Schulze PC (ed) Engineering within Ecological Constraints. National Academy Press, Washington DC, pp 111-128

Mitsch WJ, Jørgensen SE (2004) Ecological Engineering and Ecosystem Restoration. John Wiley and Sons, Inc. New Jersey

Motta R, Haudemand J-C (2000) Protective forests and silvicultural stability. An example of planning in the Aosta valley. Mt Res Dev 20:74-81

Odum HT (1971) Environment, Power and Society. Wiley Interscience, New York

Painter DJ (2003) Forty-nine shades of green: ecology and sustainability in the academic formation of engineers. Ecol Eng 20:267-273

Polster DF (2002) Soil bioengineering techniques for riparian restoration. Proceedings of the 26th Annual BC Mine Reclamation Symposium. Dawson Creek, B.C. Canada, pp 230-239

Polster DF (2003) Soil bioengineering for slope stabilization and site restoration. Paper presented Sudbury Mining and the Environment III, May 25-28, (2003) Laurentian University, Sudbury, Ontario, Canada http://www.ott.wrcc.osmre.gov/library/proceed/sudbury2003/sudbury03/122.pdf

Quine CP, Coutts MP, Gardiner BA, Pyatt DG (1995) Forests and Wind: Management to Minimise Damage. Bulletin 114. HMSO, London

Schiechtl HM (1980) Bioengineering for Land Reclamation and Conservation. University of Alberta Press, Edmonton, Alberta, Canada

Schiechtl HM, Stern R (1996) Ground Bioengineering Techniques for Slope Protection and Erosion Control. Blackwell Science Ltd, London

Sotir RB (2002) Integration of soil bioengineering techniques. Proceedings of 33rd International Erosion Control Association Conference, IECA, Orlando, Florida, pp 191-200

Steele DP, MacNeil DJ, McMahon W, Barker DH (2004) The use of live willow poles for stabilising highway slopes. TRL Report 619, Crowthorne, TRL Limited

Stokes A, Mickovski SB, Thomas BR (2004) Eco-engineering for the long-term protection of unstable slopes in Europe: developing management strategies for use in legislation. In: Lacerda W, Ehrlich W, Fontoura M, Sayao, SAB (eds) IX International Society of Landslides conference, 2004, Rio de Janeiro, Brazil. Landslides: evaluation and stabilisation, AA Balkema Publishers, vol 2, pp 1685-1690

Stokes A, Spanos I, Norris JE, Cammeraat LH (eds) 2007. Eco- and Ground Bio-Engineering: The Use of Vegetation to Improve Slope Stability. Proceedings of the First International Conference on Eco-engineering 13-17 September 2004, Thessaloniki, Greece. Developments in Plant and Soil Sciences vol. 103, Springer, Dordrecht. ISBN-10: 1-4020-5592-7; ISBN-13: 978-1-4020-5592-8.

Straskraba M (1993) Ecotechnology as a new means for environmental management. Ecol Eng 2:311-331

Chapter 2

AN INTRODUCTION TO TYPES OF VEGETATED SLOPES

Joanne E. Norris[1,2], John R. Greenwood[2]
[1] Halcrow Group Limited, Endeavour House, Forder Way, Cygnet Park, Hampton, Peterborough, PE7 8GX, U.K., [2] School of Architecture, Design and Built Environment, Nottingham Trent University, Burton Street, Nottingham, NG1 4BU, U.K.

Abstract: *Many different types of natural and artificial slopes exist throughout the world, those that have the potential and suitability for stabilizing by vegetation include earthworks on transport infrastructure, forested and agricultural slopes. This chapter introduces the reader to the different types of natural and artificial slopes and breifly discusses the potential for stabilizing each type of slope with vegetation.*

Key words: earthworks, embankments, cut-slopes, terraces, vegetation

1. INTRODUCTION

Slopes are common features of the world around us, whether they are of a geological, geomorphological or human origin. In most instances, slopes are naturally unstable unless they have been stabilized through geological time. Unstable slopes create numerous management and engineering issues as we try to maintain order and prevent slope failures from affecting our transport infrastructure, leisure activities and human life. It is hoped that by careful planning and consideration, vegetation, as an ecotechnological solution, can assist in preventing slope failures. Different types of natural and artificial slopes exist (Figure 2.1), and those which are suitable for stabilizing by vegetation include earthworks on transport infrastructure, forested and agricultural slopes (Figure 2.2). The potential for stabilizing each type of slope with vegetation is discussed below.

J.E. Norris et al. (eds.), Slope stability and erosion control: Ecotechnological solutions, 9–15.

2. NATURAL SLOPES

Natural slopes (Figure 2.2) are formed usually over long periods of time, through many geological and geomorphological processes, e.g., mountain building, glacial activity, tidal and river activity. These slopes are only stable if the soil has sufficient strength to resist the gravitational forces on the potential sliding mass. Changes in pore water pressure conditions, slope geometry or engineering works may cause these natural slopes to fail (see Chapter 3). Failure planes are e.g., rotational, translational or complex, and occur at varying depths according to the different ground conditions present.

Vegetation is unlikely to have a significant impact on slope stability where slip planes are deep-seated, due to the shallow rooting nature of many species. However, vegetation may protect the ground surface from erosion by wind and water and prevent erosion at the toe of slopes where the slope is being undercut by wave action in water courses. The stability of the toe of a slope, stabilized by vegetation, may be sufficient to maintain the stability of the slope as a whole (Coppin and Richards 1990; Gray and Sotir 1996).

Hillsides and valley slopes in rural areas are commonly planted with woodlands and managed forests. In these particular areas, individual tree instability due to storms and gales (see Chapter 5), rockfall (see Chapter 7)

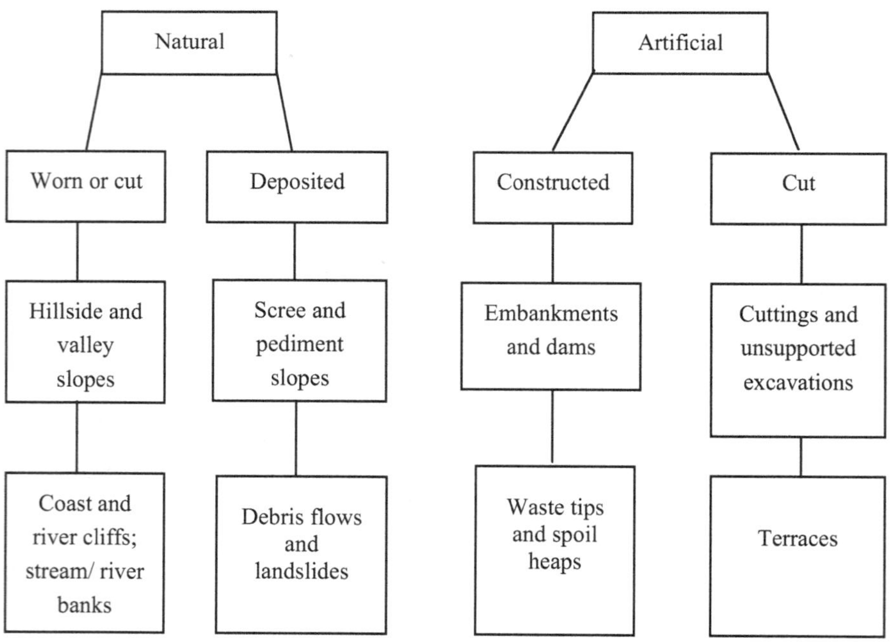

Figure 2-1. Different types of natural and artificial slopes (after Whitlow 2000).

Earthworks on transport infrastructure - railway embankment (Photo: J.E. Norris)

Highway cut-slope (Photo: J.E. Norris)

Forested slope (Photo: M. Genet)

Natural slope (Photo: L.H. Cammeraat)

Terraced slopes (Photo: Y. Chen)

Abandoned bench terrace (Photo: R. van Beek)

Figure 2-2. Examples of artificial and natural slopes.

or debris flows may be more of a problem than slope stability. Deforestation and wildfires on these types of slopes may also lead to increased soil erosion.

Many drainage channels exist on hillslopes and in valleys. The streams and rivers that meander and flow down these slopes may undercut the hillslopes and cause bank instability. Ground bio-engineering is an accepted engineering technique for stabilizing bank erosion and instability caused by fast flowing water, and as such is not specifically covered in this book. The reader is therefore referred to published texts for advice on river bank stability (e.g., Schiechtl 1980; Gray and Sotir 1996; Schiechtl and Stern 1996, 2000).

Eco-engineering methods are particularly suited to natural slopes, where management is generally long-term and the site is large-scale.

3. ARTIFICIAL SLOPES OR EARTHWORKS

Artificial slopes or earthworks are either cut into natural rock or soil or built up to form embankments, dams, waste tips or spoil heaps. Vegetation could be used for stabilizing cut slopes in soil, soil embankments, waste tips, spoil heaps and terraced slopes. It is less likely to be of value in dams where engineering stability is critical and vegetation could affect soil permeability. Ground bioengineering methods are commonly used on artificial and terraced slopes, as this fast and effective solution can be considered during slope construction and remediation.

3.1 Embankments

Embankments typically occur along highways, railways and canals (Figure 2.2) and are made from materials such as soil or rock excavated from elsewhere and placed on natural ground. The changes in condition of these materials with time and rate of deformation have critical influences on the safe and efficient operation of the transport system. Embankment stability is dependent on soil material; presence of water; shrink and swell cycles induced by seasonal moisture changes and vegetation; slope geometry, angle and height; construction method and type of foundation, and age. External factors such as vandalism, erosion and burrowing animals can cause loss of embankment performance (Perry et al. 2003a).

Slope failure can either be in the form of small-scale shallow translational slides, where the failure is contained entirely within the embankment side slopes and maximum depth of rupture does not exceed 2.0 m, or deep rotational slips that run from the crest through the embankment to the underlying foundation material to emerge beyond the toe. The type of slope failure is different for each transport sector due to the variation in construction methods, soil materials, drainage provision and function. Slope failure in embankments during and after construction is sometimes associated with the interface between the natural ground and the fill material. Pore water pressures and seepage within the embankment and natural ground may exacerbate slope failures. Where the original topsoil was left in place, a potential rupture surface may be formed (Coppin and Richards 1990; Greenwood et al. 2001; Perry et al. 2003a).

A suitable combination of vegetation types, e.g., shrubs and trees, and ground bio-engineering solutions, e.g., willow poles, can help to stabilize embankments that may be prone to the shallow translational slide failure (Coppin and Richards 1990; MacNeil et al. 2001; Marriott et al. 2001; Operstein and Frydman 2000; Steele et al. 2004; Norris 2005). Vegetation may help to stabilise the toe of deeper slips but generally deep rotational slips at depths greater than 2.0-3.0 m would be out of the zone of influence of many tree roots. For deep-seated slides, a combination of geosynthetics and vegetation may be more appropriate.

3.2 Cut-slopes and cuttings

Infrastructure cuttings and cut-slopes (Figure 2.2) are excavations in existing ground with side slopes and a trafficked surface, providing passage for road, rail and canal traffic across natural ground to maintain vertical alignment. The change in condition of the soils with time and the rate of deformation of the cutting again affect the safe and efficient use of the transport corridor (MacNeil et al. 2001; Marriott et al. 2001; Perry et al. 2003b).

The stability of a cut-slope can be affected by a reduction in the strength or stiffness of the soil through which the cut is made; a change in the external disturbing static and dynamic forces acting on the soil structure; change in geometry and the presence of water. Slope failures on cut-slopes occur in a similar manner to failures on embankments, therefore, the application of vegetation on cut slopes may be applied in the same way as for embankments, i.e., by using a combination of vegetation types to intercept shallow translational failures and by placing vegetation at the toe of slopes.

3.3 Terraces

Terraced slopes (Figure 2.2) are common features in many parts of Asia (Storey 2002) as well as Mediterranean regions, built to conserve soil and water on steep slopes for a variety of agricultural uses. However, if traditional methods are used but not implemented correctly through lack of training, care or resources, soil loss can be rapidly increased. For example, if hill terraces for the cultivation of crops are poorly constructed or maintained, topsoil erosion and slope instability will be exacerbated through water collecting on oversteepened terraces (Sidle et al. 2006). If the terraces collapse, breaches will focus surface runoff leading to gully formation and increased sediment transport downslope (McConchie and Ma 2002). Furthermore, changes in agricultural practice have led to wide scale abandonment of terraced slopes. Abandonment of terraces can result in the loss of vegetation and root reinforcement thus leading to an increase in the rate of soil erosion (Goudie 2000; Cammeraat et al. 2005; van Beek et al. 2005).

4. REFERENCES

Cammeraat E, van Beek R, Kooijman A (2005) Vegetation succession and its consequences for slope stability in SE Spain. Plant Soil 278:135-147

Coppin NJ, Richards IJ (1990) Use of Vegetation in Civil Engineering. CIRIA, Butterworths, London

Gray DH, Sotir RB (1996) Biotechnical and Soil Bioengineering Slope Stabilization: A Practical Guide for Erosion Control. Wiley & Sons, Inc., New York

Greenwood JR, Vickers AW, Morgan RPC, Coppin NJ, Norris JE (2001) Bioengineering – the Longham Wood Cutting field trial. CIRIA PR 81, London

Goudie A (2000) The Human Impact on the Natural Environment. 5th edn. Blackwell, Oxford

MacNeil DJ, Steele DP, McMahon W, Carder DR (2001) Vegetation for slope stability. TRL Report 515, TRL Limited, Crowthorne

Marriott CA, Hood K, Crabtree JR, MacNeil DJ (2001) Establishment of vegetation for slope stability. TRL Report 506, TRL Limited, Crowthorne

McConchie JA, Ma H (2002) A discussion of the risks and benefits of using rock terracing to limit soil erosion in Guizhou Province. J Forest Research (Harbin) 13:41-47

Norris JE (2005) Root reinforcement by hawthorn and oak roots on a highway cut-slope in Southern England. Plant Soil 278:43-53

Operstein V, Frydman S (2000) The influence of vegetation on soil strength. Ground Improvement 4:81-89

Perry J, Pedley M, Reid M (2003a) Infrastructure embankments – condition, appraisal and remedial treatment, C592, 2nd edn, CIRIA, London

Perry J, Pedley M, Brady K (2003b) Infrastructure cuttings: condition appraisal and remedial treatment. C591, CIRIA, London

Schiechtl HM (1980) Bioengineering for Land Reclamation and Conservation. University of Alberta Press, Edmonton, Alberta, Canada, 404 pp

Schiechtl HM, Stern R (1996) Ground Bioengineering Techniques for Slope Protection and Erosion Control. Blackwell Science Ltd, London

Schiechtl HM, Stern R (2000) Water Bioengineering Techniques: For Watercourse Bank and Shoreline Protection. Blackwell Science, Oxford

Sidle RC, Ziegler AD, Negishi JN, Rahim Nik A, Siew R, Turkelboom F (2006) Erosion processes in steep terrain – Truths, myths, and uncertainties related to forest management in southeast Asia. For Ecol Manag 224:199-225

Steele DP, MacNeil DJ, McMahon W, Barker DH (2004) The use of live willow poles for stabilising highway slopes. TRL Report 619, TRL Limited, Crowthorne

Storey PJ (2002) The Conservation and Improvement of Sloping Land: A Manual of Soil and Water Conservation and Soil Improvement on Sloping Land. Volume 1. Science Publishers, Enfield, New Hampshire, USA

Whitlow R (2000) Basic Soil Mechanics. 4th edn, Prentice Hall

van Beek LPH, Wint J, Cammeraat LH, Edwards JP (2005) Observation and simulation of root reinforcement on abandoned Mediterranean slopes. Plant Soil 278:55-74

Chapter 3

HILLSLOPE PROCESSES: MASS WASTING, SLOPE STABILITY AND EROSION

Rens van Beek[1,2], Erik Cammeraat[2], Vicente Andreu[3], Slobodan B. Mickovski[4], Luuk Dorren[5]

[1]Utrecht Centre of Geosciences, Utrecht University, PO BOX 80.115, NL-3508 TC Utrecht, The Netherlands, [2]IBED-Physical Geography, University of Amsterdam, Nieuwe Achtergracht 166, 1018 WV Amsterdam, The Netherlands, [3]CSIC-CIDE, Camí de la Marjal, s/n. Apartado Oficial, 46470 – Albal, Valencia, Spain, [4]Jacobs UK Ltd., Glasgow G2 7HX, U.K., [5]Cemagref Grenoble, 2 rue de la Papeterie, BP 76, 38402 St. Martin d'Hères cedex, France

Abstract: *This chapter describes the dominant types of processes present on hillslopes where both gravity and running water are active. The impact of natural hillslope processes is important and is currently strongly influenced by human activity due to land use change and vegetation removal, and is becoming even greater due to climate change. Both the fundamentals of erosion and slope stability are discussed in this chapter with respect to processes, causes and impacts. To fully appreciate the role of vegetation in the remediation of adverse slope processes, the fundamentals of these slope processes are addressed. In the first part, the role of mass movements is discussed. The definitions used and physical principles underlying mass movements are explained and keys and diagnostic parameters are given to explain how to recognize certain types of mass movements in the field. The causes of mass movement are described, amongst which deforestation, adverse hydrological conditions or slope undercutting, are summarized. The main types of mass movements i.e. falls, slides and flows are then separately discussed, giving full details with regard to their causes, processes and consequences, as well as a first glimpse to the solutions to slope failure problems, which will be addressed in more detail elsewhere in the book. The second part addresses erosion processes. Accelerated erosion is considered as one of the greatest problems of land degradation as it removes the fertile topsoil at high rates. Mankind, who is removing the original vegetation for agricultural purposes, is causing this problem. Again the general principles behind soil erosion are illustrated, giving attention to the causes and the different soil erosion processes such as sheet erosion, rill and gully erosion, piping and tunnel erosion as well as tillage erosion.*

Key words: mass wasting, slope stability, soil erosion, rockfall, slides, slumps, flows

J.E. Norris et al. (eds.), Slope stability and erosion control: Ecotechnological solutions, 17–64.

1. INTRODUCTION

This chapter describes the processes involved in the transport of material over hillslopes. Hillslopes provide the gradients enabling material to be transported from the slopes themselves towards the valley bottoms, directly by gravity alone, or by water flowing down over the surface. Gravity has the potential to transfer material downslope if the material resistance to counteact it is insufficient. Similarly, water and wind flowing along the surface exert a drag on soil particles and have the potential to entrain material. After the gradient has fallen below a critical threshold to keep the material in transport, deposition occurs.

Conventionally, slope processes in which gravity alone is the dominant transporting agent are called mass movements (Brunsden 1984). Processes in which other agents dominate are called erosion, e.g. wind or water erosion. Flowing water is an important transporting agent on which the emphasis is placed here.

Although mass wasting is a natural process its incidence and impact may be exacerbated by human activities (Crozier 1986; Morgan 2005). A situation may ultimately arise in which human interests become unsafe or unsustainable. This impact is not only directly felt in the areas where material is removed or deposited and indirectly mass wasting may have an effect on soil and water quality in areas located further downstream.

In the following sections a brief overview is given of the mechanisms, morphology, causes and consequences of mass movements and erosion.

1.1 Human interaction

The role of man in triggering slope processes is considerable. The continuous expansion of agricultural, industrial and built up areas, as well as the continuous enlargement of infrastructures such as roads and railways, create new areas which are destabilized by human action, including:

- Deforestation – Removal of forests is a major issue in many countries and soils may become destabilized or prone to erosion after the removal of vegetation (Sidle and Dhakal 2002). Overgrazing also reduces vegetation cover, increasing the risks for soil degradation.
- Construction activities – Built up areas are also expanding into steeper terrain in areas with high slope failure risk. Furthermore, built up areas have high runoff, increasing the risk of floods and erosion.

- The expansion of road and train networks – By expanding these network systems through hilly terrain, considerable slope cuttings may be needed, which in turn requires significant efforts to reduce the risk of hillslope processes.
- Climate change – Although some researchers claim that the cause of climate change is still controversial, it is undoubtedly occurring at an unusually fast rate. An increase in the global temperature will result in more extreme weather events e.g., increased rainfall which in turn can trigger landslides and exacerbate surface erosion (Sidle and Dhakal 2002).

Soil erosion is particularly accelerated in many areas of the world and directly influences the food security of mankind as more and more land is needed for the production of crops.

1.2 Impacts of slope processes

The impact of mass movements and soil erosion can be dramatic. Mass movements often affect large parts of a slope at relatively fast rates, depending on the type of movement. Large and deep mass movements are very difficult to manage and in most cases cannot be stopped. In some ideal cases, mass movements can be controlled in such a way that it becomes less harmful when compared to the 'natural' situation. Controlling such movements can be done by trying to reduce the inflow of water into the mass movement area or by specific measures to relocate rivers and streams preventing increased erosion (Rupke et al. 1988). Mass movements only occur on hillslopes, whether they be artificial or natural and always deliver loose material to the toe and lower slopes, which may cause off-site effects with regards to sediment delivery to lower areas via river flow transport.

Soil erosion occurs on hillslopes and removes the fertile topsoil. When this occurs at higher rates than soil formation and weathering this loss is irreversible. It not only leads to the local formation of rills, gullies or tunnels, thus damaging agricultural fields, but also leads to considerable off-site effects such as sedimentation in valley bottoms, where it can result in blockage of roads and damage to property. Soil erosion may also lead to the siltation of reservoirs, which is of major concern. Many examples are known where reservoirs have been filled within 10-30 years after their construction by sediment produced in the uplands.

2. MASS MOVEMENTS

2.1 Introduction: Terminology, General Principles and Recognition

2.1.1 Terminology

The term mass movements is used here (Brunsden 1984) as a more generic term for those processes that Varnes (1978) called landslides and defined as "a downward and outward movement of slope forming material under the influence of gravity". Slope instability is used to describe the resulting deformation of the slope and the term failure the onset of movement. Mass movements encompass a wide range of slope deformations associated with slope instability. In addition to sliding along a discrete shear plane they include the free, downslope movements of rocks and rock masses, (falls and topples), the latter exhibiting a rotational component, and flows. In this book, emphasis is placed on those mass movements of which the occurrence or behaviour is influenced by vegetation. Some large-scale deformations e.g., lateral spreading, cambering and sagging are therefore ignored or only briefly mentioned (Varnes 1978; Hutchinson 1988).

2.1.2 General principles of slope instability

For the prediction and remediation of mass movements it is essential that the stability and likely deformation mechanisms of a slope are understood.

Slope stability depends on the equilibrium between the driving and restoring forces that act on a potentially unstable soil mass. The driving forces acting on slope material, including gravity, result in a shear stress, τ, that must be counteracted by the available shear strength. This concept forms the basis of the safety factor, FOS, which is the ratio of the maximum available shear strength over the shear stress. If this ratio is larger than one, the slope can be considered stable (see Chapter 5).

The Mohr-Coulomb failure criterion describes the available shear strength of rocks and soils adequately in most cases and is the most widely used constitutive equation of shear strength. This criterion attributes the shear strength of a material to a finite cohesion and a frictional component. Cohesion is expressed as a stress and can be interpreted as the total of attractive forces between particles per representative bulk area e.g., 1.0 m^2 of material, of the shear plane along which the shear strength is mobilised. This stress is a major constituent of the strength of *plastic* or fine-grained soils such as clays and silts. The frictional resistance is mobilised at the particle contacts and increases with particle size (Table 3-1). The frictional component is proportional to the inter-particle forces, that is represented by the normal stress acting on the representative bulk area, σ′. At failure, the maximum available shear strength is mobilised which can be expressed by (Lambe & Whitman 1979):

$$\tau_f = c' + \sigma' \tan \phi' \tag{1}$$

where τ_f is the shear strength at failure, c′ is the cohesion, σ′ is the normal stress (all in units of stress) and ϕ′ is the angle of internal friction. Figure 3.1 represents Equation [1] graphically.

Table 3-1. Mohr-Coulomb shear strength of different materials at peak strength.

Material	Shear strength		Source
	c′ (kPa)	ϕ′	
Plastic (cohesive) fine-grained soils: clays	6-10	17-24°	Ortiz et al. (1986)
Plastic (cohesive) fine-grained soils: silts	≈ 3	≈ 25°	Ortiz et al. (1986)
Granular (frictional) coarse soils: loose sands	≈ 0	≈ 32°	Ortiz et al. (1986)
Granular (frictional) coarse soils: dense sands and gravel	≈ 0	≈ 35°	Ortiz et al. (1986)
Weak rock: heavily fractured or poorly consolidated	≈ 38	≈ 14°	Goodman (1980)
Competent rock: intact and sound material	6-66	27-55°	Goodman (1980)

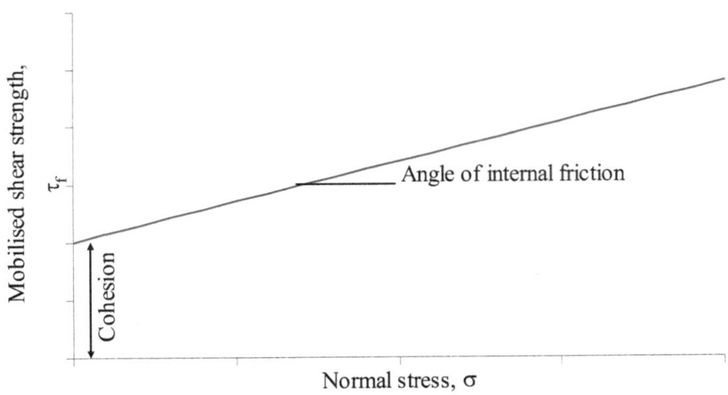

Figure 3-1. Mohr-Coulomb failure envelope.

The primed variables of Equation 1 signify that shear strength is expressed in terms of *effective stress* (Terzaghi's principle; Lambe and Whitman 1979). When pore pressures are present, for example below the water table, they carry part of the inter-particle stress and the total normal stress, σ, is reduced by the pore pressure u to the effective normal stress σ'.

$$\sigma' = \sigma - u \tag{2}$$

The effective shear strength is called the *drained shear strength* as it is determined at strain rates that are sufficiently low to allow complete drainage and avoid the negative effect of pore pressures on the shear strength. Excessive pore pressures can be expected in an engineering context as a result of rapid loading or draw-down. In these cases, it is more appropriate to work with the *undrained shear strength* and in terms of total stresses. In such cases, the contribution of the frictional component will be virtually nil ($\phi = 0$ analysis). Only when excess pore pressures have dissipated and the fabric of particles carries all loads, is it appropriate to use the drained shear strength again. For this reason, the undrained and drained shear strengths are considered to be characteristic for the short- and long-term stability of a slope respectively (Skempton 1964).

Failure upsets the soil fabric and changes the shear strength accordingly. Dense granular soils, e.g., sands, often dilate when the interlocking particles are moved over each other and the frictional resistance decreases. Likewise, large displacements in a concentrated shear zone destroy the cohesive bonds between particles. Consequently stress-strain graphs often exhibit a drop in the shear strength after a peak at failure and trail off to a residual value at large strains (Figures 3.2 and 3.3). This residual shear strength should be considered in the case of reactivation whereas the peak shear strength is appropriate in the case of first-time failures only.

Some rock and soil materials can be highly problematic with regards to slope stability. Swelling clays, e.g., smectites, can expand and upset the balance of a slope or act as lubricant in joint systems. Likewise, some volcanic derived soils containing amorphous Al-silicates experience a substantial loss in shear strength upon wetting. Some rock types such as gypsum, salt and limestone are prone to dissolution, which may threaten the integrity of the rock mass as a whole (Seijmonsbergen 1992). Changes in the soil fabric after failure can also alter the available pore space. This is often the case in loosely packed materials such as loess or peat deposits. Upon contraction, excessive pore pressures may form by the compression of water or air and force the material to behave as a viscous fluid that can sustain less shear stress. This compression affects stability negatively and may result in much larger displacements and velocities. Dilation of the shear zone can increase the pore space and exert a suction that increases the inter-particle stresses (*viscous drag*; Nieuwenhuis 1991). Such a phenomenon attenuates slope movement, in particular in the case of large landslides in fine-grained soils. It is important to realise, therefore, that the shear strength counteracting slope instability is not constant over time; material is generally able to mobilise more strength to ward off first-time failure than to prevent reactivation.

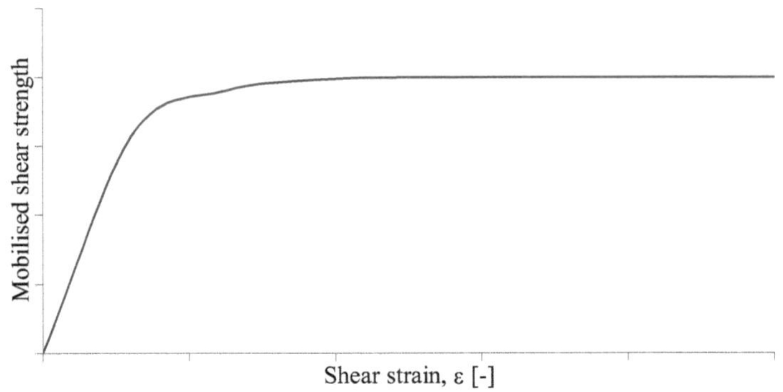

Figure 3-2. Material with constant stress-strain behaviour.

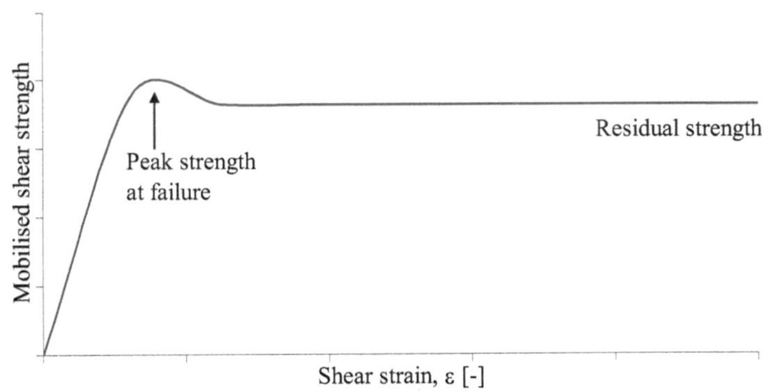

Figure 3-3. Material exhibiting strain-softening.

2.1.3 Recognition of mass movement types

It is imperative that due consideration should be given to the hazard of slope instability before any activities are deployed or engineering works carried out in hilly terrain. The recognition of those areas that are prone to failure or areas that have been subjected to slope instability in the past or present must be performed to avoid inadvertent development on a site.

Site characteristics provide crucial information about the hazard of potential slope instability. Because of the complex causes of most landslides, it is hard to give precise criteria. Therefore, the site characteristics listed in Table 3-2 only provide guidance to recognise potentially or actually unstable terrain (compiled from Crozier 1984; Sidle et al. 1985; Cooke and Doornkamp 1990; Rib and Liang 1978; Cruden and Varnes 1996; Dikau et al. 1996a).

Site characteristics can help to distinguish active from inactive mass movements (Table 3-3). Active mass movements are defined here as those that have shown movement in the recent past and can be expected to be reactivated in a foreseeable period. This potential hazard for reactivation is central to the sustainability of certain activities or the desirability of engineering solutions.

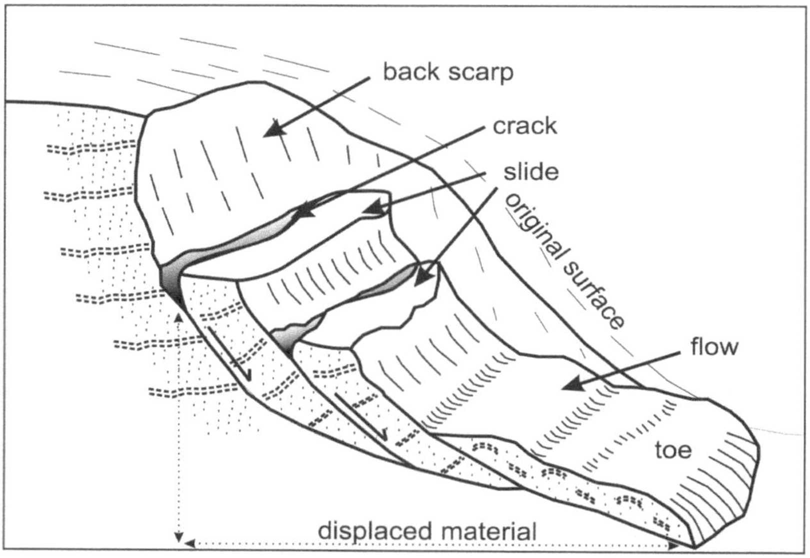

Figure 3-4. Mass movement terminology (after Summerfield 1991).

Mass movement types can be recognised on the basis of the characteristics of the different mass movement parts (Figure 3.4). A simplified scheme for the recognition of mass movements is given here based on the original of Rib and Liang (1978; see Table 3-4).

Care is required with the interpretation of site characteristics. Undoubtedly, any judgment on mass movement hazards will be subjective and it is strongly advised that local expertise is consulted, as distinct conditions may be important for the initiation and reactivation of mass movements in a given region.

Table 3-2. Site characteristics of slopes prone to instability.

Site characteristic	
Morphology	
Gradient	Moderately steep for landslides (>10°) to extremely steep for rockfalls (>35°). Some flows can maintain momentum even on very gentle slopes.
Shape	Convergent or irregular in profile.
Height	Short steep slopes for rotational slides, long slopes for translational slides.
Material	
Slope material	Plastic soils, material sensitive to physical or chemical weathering or heavily fractured or jointed rock.
Stratigraphy	Alternation of weaker and stronger beds, of different permeability.
Hydrology	Signs of ponding and springs, presence of gleyic horizons indicating stagnating water in the soil.
Drainage	Heavily dissected by ephemeral or permanent streams with signs of undercutting at the base of the slope or signs of disrupted drainage.
Climate	Periods of intense or prolonged rainfall or rapid snowmelt; strong diurnal and seasonal variations in temperature, e.g. freeze-thaw.
Seismicity	Evidence of moderately strong to strong earthquakes.
Past activity	Signs of previous slope movements (creep, sliding) and/or surface wash.
Vegetation	Irregular stands and/or deformed or underdeveloped vegetation; exposure of roots in cracks or at the surface.
Human activity	Evidence of poor site management (leakage of sewer systems, blocked drains etc.) or extensive changes to the shape or composition of a slope. On a marginally stable slope, human intervention can easily upset the critical balance.

Table 3-3. Distinct features of active and inactive mass movement (Crozier 1984).

Active	Inactive
• Scarps, terraces and crevices with sharp edges; • Crevices and depressions without secondary infilling; • Secondary mass movement on scarp faces;	• Scarps, terraces and crevices with rounded edges; • Crevices and depressions infilled with secondary deposits;

• Surface-of-rupture near marginal shear planes show fresh slickensides and striations; • Fresh fractured surfaces on blocks; • Disarranged drainage system; many ponds and un-drained depressions; • Pressure ridges in contact with slide margin; • No soil development on exposed surface-of-rupture; • Presence of fast-growing vegetation species; • Distinct vegetation differences on and off slide; • Tilted trees with no vertical growth; • No new supportive, secondary tissue on trunks.	• No secondary mass movement on scarp faces; • Surface-of-rupture near marginal shear planes show old or no slickensides and striations; • Weathering on fractured surfaces of blocks; • Integrated drainage system; • Marginal fissures and abandoned levées; • Soil development on exposed surface-of-rupture; • Presence of slow-growing vegetation species; • No distinct vegetation differences on and off slide; • Tilted trees with new vertical growth above inclined trunk; • New supportive, secondary tissue on trunks.

2.2 Causes of mass movements

"The processes involved in slope movements comprise a continuous series from cause to effect" (Varnes 1978). It is therefore often difficult to attribute slope instability to a single factor (Bogaard 2001). Clearly some factors are more dynamic than others, which can be considered quasi-static on human timescales. Therefore, the spatial distribution of these least changeable factors determines the *susceptibility* of a slope or a set of slopes to failure (e.g., geology, slope gradient, slope aspect, elevation, soil properties, and long-term drainage patterns; Dai and Lee 2001). Given this susceptibility, the more dynamic factors such as rainfall or seismic events trigger the instability. Based on this distinction, Crozier (1986) proposed therefore a distinction in preparatory factors that increase the susceptibility of a slope to failure over time and triggering factors that upset the balance momentarily.

The frequency of potential triggers defines the incidence of mass movement (Van Asch and Van Steijn 1991; Crozier 1986). Consequently, mass movement hazard consists of a spatial and a temporal component that needs to be evaluated jointly (Varnes 1984) and the resulting mass movements pose a risk to activities, both in its source area and along its track. After instability has occurred, the resulting mass movement may remain active for a long time, which poses a further risk for any activities in the affected area.

Table 3-4. Summary of terrain characteristics that assist in the recognition of the mass.

Landslide type	Crown	Main scarp	Flanks	Head	Body	Foot	Toe
Fall, topple	Consists of loose rock, debris or soil; probably has cracks behind scarp; in rock has irregular shape controlled by local joint system.	Usually almost vertical, irregular, bare and fresh, consisting of joint or fault shears in rock and spalling on the surface if debris or soil.	Mostly bare edges of rock, often nearly vertical.	Usually not well defined; consists of fallen material that forms a heap of rock, debris or soil next to the scarp.	Fall: irregular surface of jumbled rock that slopes away from the scarp and that, if large, trees may show direction of movement radial from the scarp; may contain depressions. Topple: consists of unit or units tilted away from the crown.	Commonly buried; if visible, generally shows evidence of reason for failure, such as prominent joint or bedding surface, underlying weak rock, or banks undercut by water.	Irregular piles of debris or talus if slide is small; may have rounded outline and consist of broad, curved transverse ridge if slide is large.
Rotational slide (single, multiple or successive)	Has cracks that tend to follow fracture patterns in the original rock; in debris or soil cracks are mostly curved concave toward the slide.	Steep, bare, concave toward the slide and commonly high; may show striae and furrows on the surface running from crown to head; may be vertical in the upper part.	Have striae with strong vertical components near head and strong horizontal component near foot; have scarp height that decreases toward foot; may be higher than original ground surface between foot and toe; have 'en echelon' cracks that outline slide in earlier stages.	Remnants of land surface flatter than original slope or even tilted into hill, creating at base of main scarp depressions in which perimeter ponds form; has transverse cracks, minor scarps, grabens, fault blocks; bedding attitude different from surroundings; trees lean uphill.	Consists of original slump blocks generally broken into smaller masses; has longitudinal cracks, pressure ridges and occasional over thrusting; commonly develops small pond just above the foot.	Commonly transverse cracks developing over the foot line and transverse pressure ridges developing below the foot line; has zone of uplift, no large individual blocks and trees that lean downhill.	Often a zone of earth flow of lobate form in which material is rolled over and buried; has trees that lie flat or at various angles and are mixed into the toe material; in rock there is little or no flow, often nearly straight and close to the foot may have steep front.

Translational slide: slab or block slide	Has cracks most of which are nearly vertical and tend to follow the contour of the slope.	Is nearly vertical in the upper part and nearly planar and gently sloping in the lower part.	Low scarps with vertical cracks that usually diverge downhill.	Relatively undisturbed and has no rotation.	Usually composed of single or few units; is undisturbed except for common tension cracks that show little or no vertical displacement.		Flows or overrides ground surface.
Translational slide: rock slide, debris slide, mudslide	Consists of loose material and has cracks between blocks.	Usually stepped according to spacing of joints or bedding planes in rock; has irregular surface in upper part and is planar or gently sloping in lower part.	Irregular.	Many blocks of rock, debris or soil.	Rough surface of many blocks some of which may be in approximately their original altitude but lower if movement was slow; shows flow structure.		Consists of an accumulation zone of rock, debris or soil; spreading and lobate often consists of material rolled over and buried.
Debris flow	Few cracks	Typically has serrated or V-shaped upper part; is long and narrow, bare and commonly striated.	Commonly diverges in direction of movement.		Consists of large blocks pushed along in a matrix of finer material; has flow lines; follows drainage patterns; is very long compared to its breadth.	Buried in debris.	Spreads laterally in lobes; if dry, may have a steep front about a meter high.
Soil flow	Few cracks	Steep and concave toward slide; may have a variety of shapes in outline; nearly straight, arcuate, circular or bottle-shaped.			Conical heap of soil, equal in volume to the head region.		Spreading and lobate.

Mass movement hazard equally applies to natural and man-made slopes. However, tolerances of mass movement occurrence may vary widely between slopes as a function of the vulnerability of the elements at risk.

Based on the safety factor concept the causes of slope instability can be subdivided into internal and external causes (Chandler 1986; Gostelow 1996). Internal causes reduce the available resistance of the soil whereas external causes increase the disturbing forces acting on the soil mass (Table 3-5).

Table 3-5. List of examples of mass movement causes compiled from Varnes (1978); Crozier (1984); Hutchinson (1988); Cruden and Varnes (1996) and Wieczorek (1996).

Internal	
Changes in water regime	Pore pressure increase or matric suction decrease upon wetting by rainfall, snow melt or leakage from utilities
Weathering, erosion and progressive failure	Deterioration of cohesion and cementation bonds Freeze/thaw cycle Shrink/swell cycle Seepage erosion
External	
Loss of support	Slope erosion, riverbank erosion, wave erosion, glacial and stream incision Excavation, mining Draw-down of reservoir levels
Increased surcharge	Vegetation growth Increasing weight because of wetting Accumulation of sediment Landfill Building

Although mass movements are natural processes, their incidence and impact may be exacerbated by human activities (Crozier 1986). In particular, land use changes play an important role as they can affect large areas over relatively short time spans and mechanical and hydrological properties of vegetation also effect slope stability (Sidle et al. 1985; Coppin and Richards 1990; van Beek et al. 2005). Such land use changes can affect large areas over relatively short time spans and may lead to profound changes in mass movement activity (Van Beek and Van Asch 2004). Possible positive effects of land use change on stability are increased root reinforcement and attenuation of pore pressures by increased interception and transpiration, if vegetation cover and biomass increase (see Chapter 4). However, increased infiltration rates partly cancel out the positive hydrological effects under high rainfall totals. Negative effects of land use change occur after clearing of the vegetation when root reinforcement is lost or by irrigation when slope material softens and pore pressures are elevated e.g., after irrigation soil slips have occurred around the Hei Fan Tai loess plateau, PR China (Dijkstra et al. 2000; Figure 3.5).

Figure 3-5. Failures along the margin of the Hei Fan Tai Plateau (Photo: T. Dijkstra).

2.3 Processes of slope instability

Several classifications of mass movement processes exist of which the most well-known are those of Hutchinson (1988) and Varnes (1978). The scheme by Varnes (1978) has been adapted by the EPOCH project for the European situation (Dikau et al. 1996b; EPOCH 1993) and revised by Cruden and Varnes (1996), which has been adopted here in a simpler form. This classification distinguishes the different processes of slope deformation and three main material types (Table 3-6). Each material type possesses a different strength and post-failure behaviour (see also Table 3-1). The material types are:

- Earth: predominantly (> 80%) fine-grained soil (< 2 mm);
- Debris: contains between 20 to 80% of coarse soil material (≥ 2 mm) in a matrix of fine-grained soil;
- Rock: a hard or firm mass that was intact and at its natural place before the initiation of movement.

Rocks have a high intrinsic strength but contain discontinuities such as fissures and bedding planes that constitute planes of internal weakness along which, dependent on their orientation, displacements will preferentially take place. Moreover, they form pathways along which water and air may enter and reduce the strength of the rock mass further by physical and chemical weathering. Both earth and debris are either formed by deposition of transported material or formed in place by the weathering of rock or primary soils. Compared to rock, earth and debris contain many pores that may be filled with air and water. Some materials may resemble rock such as residual soils or be classified as such for geological reasons e.g. London Clay, but behave essentially as soils and should be dealt with accordingly (so-called *engineering soils*).

Table 3-6 summarises the mass movement processes of which the characteristics and causes are described in more detail in the following sections. These processes are distinguished on the basis of the mechanism

Table 3-6. Classification of mass movement types (Varnes 1978; Cruden and Varnes 1996; EPOCH 1993).

Type	**Rock**	**Debris**	**Earth**
Fall *Very rapid to extremely rapid*	Rock fall	Debris fall	Earth fall
Topple *Extremely slow to extremely rapid*	Rock topple	Debris topple	Earth topple
Slide: Rotational (slump) (single/multiple/successive) *Extremely slow to rapid*	Rock slump	Debris slump	Earth slump
Slide: Translational (non-rotational) *Extremely slow to rapid*	Rock block slide	Debris block slide	Earth slab slide
(planar) *Extremely slow to rapid*	Rock slide	Debris slide	Mudslide
Flow *show to extremely rapid*	Rock flow	Debris flow	Earth flow
Complex	e.g. Rock avalanche	e.g. Flow slide	e.g. Slump-earthflow

of deformation, the size and shape of the unstable mass and the overall velocity. In addition to their temporal occurrence, these characteristics determine largely the hazard that mass movements pose.

The types of mass movement in Table 3-6 are idealised representations of true mass movements. In reality, one mass movement process often transforms into another along the slope (*complex mass movements*, see Figure 3.6). Mass movements that involve different processes at the moment of failure are called *compound mass movements.*

Figure 3-6. Large complex earth flow near Trivento, Molise, Italy, including secondary slides and earth flows (Photo: E. Cammeraat).

2.3.1 Falls and Topples

Description

Falls and topples start with the detachment of material from a steep slope along a surface on which little or no shear displacement takes place. In falls, material moves by free fall, bouncing and rolling. Topples distinguish themselves from falls in that the movements pivot around the base of the slope. The differential movements that are required for toppling may arise from weaker basal strata (flexural topple), orientation (block topple) and small strains accumulated along numerous cross-joints (block flexural topples; Dikau et al. 1996c).

Falls and topples occur in all materials where sufficiently steep slopes exist. Earth and debris cannot sustain such slopes and the volume involved is generally small. Rock sustains steeper and larger slopes and greater volumes are involved. A negative relation exists between magnitude and frequency (Douglas 1980; Whalley 1984). Frequent falls and topples are associated with steep, highly fissured rock masses e.g. limestone. Repeated activity at the same location may lead to the formation of talus cones that have angles of repose close to the friction angle and show some sorting of material (Kirkby and Statham 1975; Statham and Francis 1986; Evans and Hungr 1993). Large falls generate a movement of dry, cohesionless debris that is displaced at high speeds (rock avalanche; Angeli et al. 1996). Deposits of such large, singular events are more chaotic in nature and discordant to the general topography (Flageollet and Weber 1996).

Causes

Steep slopes are a prerequisite for the occurrence of falls and topples. These kind of slopes can be found where slopes have been undercut by fluvial erosion, abraded by glacial erosion or uplifted, for example by volcanic activity (Flageollet and Weber 1996). Equally, over-steepened slopes may result from human activity that alter the slope e.g., quarrying and the construction of cut slopes. In earth and debris, most falls and topples occur in cohesive material in which tension cracks have developed or concern individual blocks that have been excavated by erosion.

In rock, discontinuities often delineate an unstable block. Tension cracks are important as they are often aligned parallel to the rock face and intersected by other sets. Tension cracks open due to decompression, for example as the result of deglaciation or unloading events. Over time, physical and chemical weathering affects the strength along these discontinuities negatively (Schumm and Chorley 1964; Day 1997). Asperities along the contact are worn down while the finer infill acts as lubrication and blocks drainage. Preparatory factors that can eventually lead to the initiation of rockfalls and topples include, among others, freeze-thaw cycles, periodic wetting leading to swell of clayey infills and dissolution/oxidation of rock-forming minerals, root wedging etc (Whalley 1984). Several short-lived phenomena can act as triggering factors (Dorren 2003): the overall balance of the slope can be upset by dynamic loads such as seismicity, vibrations due to blasting or heavy traffic and the passing of animals or humans. The block can become detached from the slope by pressures that act within the discontinuity, such as hydrostatic pressures after rapid snowmelt or intense rainstorms or due to the freezing of stagnating water in the cleft.

Movement

Prior to detachment of material in the source area, blocks may experience creep and accelerate exponentially over relatively long periods. This period may be indeterminate for slow, continuous toppling (Dikau et al. 1996c). After the rock has been detached and starts to move, it descends the slope in different modes of motion. These modes of motion strongly depend on the mean slope gradient (Figure 3.7). The three most important modes of motion are: freefall through the air, bouncing on the slope surface and rolling over the slope surface (Erismann and Abele 2001).

Freefall of rocks occurs on very steep slopes (Figure 3.8). According to Ritchie (1963) freefall occurs if the slope gradient below the potential falling rocks exceeds 76°, but in different field situations this value varies, therefore Figure 3.7 shows that around 70° the motion of the rock gradually transforms from bouncing to falling. During freefall of rocks, two different movements could occur. The first is the translation of the centre of rock and the second is rotation of the block around its centre (Azzoni et al. 1995). Translation and rotation are important, because falling rocks are hardly ever round. Following rotation in the air, a rock could bounce into a different direction after impact, compared to preceding directions. If the mean slope gradient decreases in the down slope section, a rock bounces on the slope surface, against barriers or against other falling rocks after freefalling. During the first bounce after freefalling, a rock tends to break, especially rocks with structural faults (Bozzolo and Pamini 1986). Whether a rock breaks or not, between 75% and 85% of the energy gained in the initial fall is lost in this first impact (Broilli 1974; Evans and Hungr 1993).

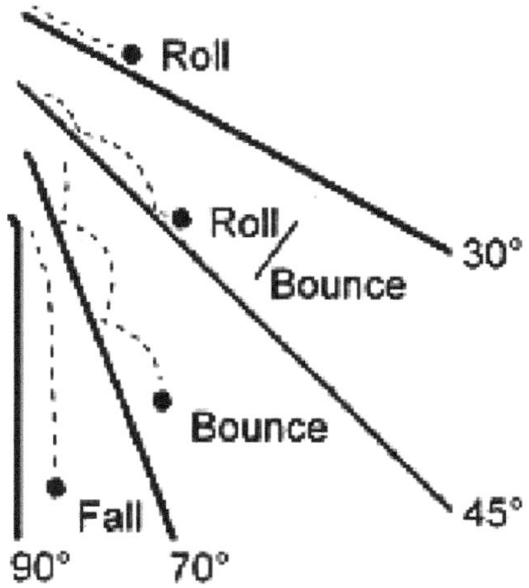

Figure 3-7. General modes of motion of rocks during their descent on slopes related to the mean slope gradients (modified from Ritchie 1963).

If the mean slope gradient is less than approximately 45°, a bouncing rock gradually transforms its motion to rolling because the rock has gathered rotational momentum during the preceding motions. A rolling rock is nearly constantly in contact with the slope surface (Hungr and Evans 1988). During the transition between bouncing and rolling, the rock rotates very fast and only the edges with the largest radius maintain contact with the slope. Thereby, the centre of gravity moves along an almost straight path, which is an effective mode of motion with respect to energy loss. In fact, this combination of rolling and short bounces is one of the most economic displacement mechanisms (Erismann 1986). Sliding is another mode of motion over the slope surface, but this generally only occurs in the initial and final stage of a rockfall. If the mean slope gradient increases, a sliding rock starts falling, bouncing or rolling. If the mean slope gradient does not change while sliding, the rock usually stops because of energy loss due to friction (Bozzolo and Pamini 1986).

Figure 3-8. Example of a large block fallen from a limestone cliff induced by undercutting and removal of underlying softer marl rocks (Rio Mula, SE Spain) (Photo: E. Cammeraat).

After going through different modes of motion, a moving rock stops. The velocity and therefore stopping of a falling rock mainly depends on the mean slope gradient, since falling rocks generally accelerate on steeper slopes and decelerate on flatter slopes. But apart from the mean slope gradient, the velocity of the falling rock also depends on the size of the rock and on the material covering the slope e.g., soil, scree and vegetation. Small rocks retard more easily than bigger rocks, firstly because during a rockfall, the total kinetic energy of small rocks is lower than that of bigger rocks, secondly large obstacles like trees could more easily stop small rocks (see Chapter 7) and thirdly, small rocks retard more easily in depressions between larger rocks on talus slopes. These are the main causes of the sorting effect on talus slopes (Kirkby and Statham 1975; Statham 1976; Statham and Francis 1986). Fine material is found near the base of the rock face and down slope the average rock size increases. Consequently, the biggest rocks are mostly found near the base of the talus slope (Evans and Hungr 1993). On alpine talus slopes, this sorting effect is neither linear nor fully exponential. Generally, the sorting effect only accounts for the upper part of the talus slope, since avalanches and debris flows deposit boulders with variable rock sizes mainly at the base of talus slopes (Jomelli and Francou 2000).

Table 3-7. Diagnostic features of falls and topples (after Rib and Liang 1978; Flageollet and Weber 1996; Dikau et al. 1996c).

Potential	Relict
1. A slope face: steep to vertical, overhanging or undercut; 2. Cracks close to the face; 3. A sufficiently large area to unload material from the source area; 4. Discontinuities form unfavourable sets projecting out (*falls*) or running parallel to the slope (*topples*); 5. Materials sufficiently strong (cohesive soils, rock) to sustain the slope over a period of time; 6. Material liable to deterioration: excavation of more resistant blocks/boulders (soil), physical or chemical weathering (rock) or worn-down or gouge-filled discontinuities; 7. The presence of weaker basal layers; 8. Environments experiencing periodic freezing and/or large water inflow (snow melt, rainstorms) or dynamic loading (blasting, seismicity).	1. Clear, near-vertical scarp exposing fresh material and showing signs of decompression, e.g. widening tension cracks. Blocks of material tilting away from the scarp (*topples*); 2. At the base of the slope or scarp, accumulation of broken material when freefall has occurred, recognisable as scree, open-work rock textures and detached boulders. Or the presence of disturbed strata in the case of slow, continuous movement; 3. At the toe of the accumulation zone, irregular piles of debris have a rounded outline and consist of broad, curved transverse ridges if volume is large and topography permits; 4. Large volumes may block valley floors with massive debris, occasionally damming streams to form reservoirs.

Stopping of rocks is an abrupt rather than a gradual process. Stopping occurs because energy is lost due to collisions and friction forces that act on the rock during transport over slope surfaces. The friction force of a moving rock is not only dependent on the rock shape, but also on the surface characteristics of the slope (Statham and Francis 1986). Slope surface characteristics might vary a lot within short distances. Therefore the friction force between a rock and the slope surface could best be characterized by a dynamic angle of friction (Kirkby and Statham 1975).

Recognition

Characteristics of falls and topples and of sites that are potentially prone to these type of mass movements are summarised in Table 3-7. See Figure 3.7 for the definition of mass movement topography.

Consequences and mitigation

Falls and topples are potentially very dangerous because of the phase of slow acceleration, the sudden collapse and the subsequent rapid displacement of large volumes of material and the erratic movement with high run-out distances of this material over the slope. Mitigation against falls and topples includes the monitoring of displacements and avoidance by exclusion or evacuation, the reinforcement of a potentially unstable slope by anchors, grouting etc. and the interception of material by catch benches and barrier fences or protection forests (Hearn et al. 1992; Peila et al. 1998; Kienholz and Mani 1994; Dorren et al. 2004, see Chapter 7).

To reduce the runout zone of falling rocks, forests can act effectively. Quantitative studies on the effect of forest cover on rockfall were carried out by amongst others Jahn (1988) and Dorren et al. (2005) and they concluded that three to ten times as many falling rocks were stopped on forested slopes compared to similar slopes without a forest cover. Zinggeler et al. (1991) also investigated the importance of trees in stopping falling rocks and concluded that topography is just as important; falling rocks lose energy by colliding with tree stems, which eventually results in stopping on flatter areas in the terrain. Hétu and Gray (2000) observed the effect of forests on scree transport on slopes. They related an increased rock concentration along forest fringes on talus slopes to an increased forest density. According to Hétu and Gray (2000), there is a constant ongoing battle between active talus slope development and forest colonization. The active front zone of the talus slope displaces downslope if a forest is disturbed by a large–scale mass movement or fire. Their study indicated that forests cannot stop the devastating effect of large magnitude rockfall events, but for low magnitude – high frequency rockfall events forests provide effective protection. Studies carried out by Dorren et al. (2005) showed that an average alpine forest reduced the rockfall hazard under an active rockfall slope with 60 – 80%. The protective effect of a forest, however, changes over time as forests are dynamic open systems. Aging of forests combined with low regeneration can result in unstable forests that provide little protection. Therefore, forest management is an essential eco-engineering technique to sustain the protective function of a forest. The specific techniques required to optimise the protective function of forests against rockfall will be described in Chapter 7.

2.3.2 Slides

Description

Sliding denotes the movement of slope material along a recognisable shear plane to which most of the movement is restricted. The shape and number of shear planes as well as the material are used to subdivide slides into:

- Rotational slides (or *slumps*) that can be either single, successive or multiple;
- Translational slides e.g. block slides or debris slides.

Varnes (1978) defines a rotational slide as a "more or less rotational movement, about an axis parallel to the slope contours involving shear displacements (sliding) along a concavely upward-curving failure surface which is visible or may be inferred". Rotational slides are subdivided into single, multiple and successive slides (Clowes and Comfort 1982; Hutchinson 1988). Single slides are one-off events whereas multiple and successive slides involve the subsequent destabilisation of an unloaded slope. Multiple slides are retrogressive and share the same basal sliding surface. Successive slides are the result of stepwise destabilisation. A further distinction of single rotational slides can be made on the basis of the position of the intersection of the shear surface with the slope; in the case of slope failures a section fails, whereas in the case of toe failures the unstable mass passes though the toe of the slope. Basal failures often occur when the mass slides over a weaker layer and mobilises a part of the base in front of the slope.

Translational slides are non-circular failures in which material moves more or less parallel to the ground surface. The shear plane is often located at a particular plane or zone of weakness. Typical examples of these phenomena are block slides in which a few units of coherent bedrock move over a gently sloping discontinuity (Ibsen et al. 1996a). Competent bedrock may also fail in wedges defined by intersecting joints or where they dip parallel to the surface (Terzaghi 1962; Goodman 1980). The scale of these features varies with the orientation and spacing of joint sets and the strength of the original or weathered material (Patton 1970).

Translational slides in loose material comprise slab slides in which weathered material slides over sound parent material (Ibsen et al. 1996b). This type of slide includes soil slips, which are shallow translational failures that affect the topsoil only. Vegetation effects have a strong influence on such shallow slips and, indeed, they often occur after logging of forests or fires in mountainous areas (O'Loughlin 1974; Cannon et al. 2001; Guthrie 2002). In coarser material, debris slides have more or less a similar appearance. Such shallow failures (0.5 to 1.5 m) extend often over a long, narrow area on steep slopes (25-45°) and concern generally colluvium, morainic drifts and strongly weathered bedrock (Sidle et al. 1985; Corominas 1996). Over steep terrain, debris slides can attain high velocities and turn into debris avalanches. Mudslides, synonymous with earth flows, are mass movements in which softened, clayey, silty or very fine sandy material moves predominantly by sliding over a discrete shear surface, often at a relative slow pace, in lobate or elongate forms (Brunsden 1984).

Causes

Slope angle is the main control of slope instability but the critical slope angle varies widely with the available shear strength, which depends primarily on the type of material. Slides occur in a wide range of materials, but rotational slides occur predominantly in thick cohesive deposits that may or may not show stratification. Slides may also occur in heavily fractured rock masses.

Translational slides are more frequent on layered soils. Shallow failures (soil slips and debris slides) occur where thin soils and drifts cover the bedrock topography. Such a lithic contact leads to higher pore pressures by impeding percolation and acts as a potential slip plane (Campbell 1975). Block slides and wedge failures occur where geologic layers or discontinuities act as planes of weakness along which the material can slide.

Slope length plays a minor role for short slopes, leading to relatively high curvatures of the slip plane and increased shearing resistance at the toe. Therefore, rotational toe or base failures are more frequent on short, steep slopes.

Processes that increase the susceptibility of a slope to failure are the removal of support, e.g., undercutting by river or sea erosion, other slope profile modification and additional static loading, especially when the slope angle exceeds the friction angle or the load is placed at the crown of a landslide. Loading and unloading, with or without the development of excessive pore pressures, are key processes in the activity of multiple and successive rotational slides and mudslides (Brunsden 1984).

Weathering may reduce the shear strength of the material or discontinuities in the long-term. Vegetation changes affect the shearing resistance over various periods. After clear-cutting or fire, surcharge losses take immediate effect. Changes to the slope hydrology and loss of reinforcement due to root deterioration take longer to come into effect (Ziemer and Swanston 1977). Progressive weakening of material from its peak to residual strength, e.g., by creep, is an important factor in the initiation of block slides and multiple and successive rotational slides. Likewise, unloading and the subsequent opening of joints may increase the weathering and susceptibility of rock slopes to failure.

The most common triggers of slides are earthquakes or other vibrations that upset the equilibrium of the slope, and also excessive or prolonged rainfall or snowmelt. Snowmelt and rainfall lead to the build-up of pore pressures that adversely affect the available shearing resistance. The typical disruption of drainage on rotational slides and the percolation of ponded water from the scarp along the slip surface can lead in turn to poor drainage and prolonged activity (Crozier 1984). In rock clefts, the available storage is small and the rise in pore pressures after snowmelt or rainfall sudden and large (Sorriso-Valvo and Gullà 1996).

Movement

According to the classification of Table 3-6, movement ranges from very slow to extremely rapid and the variations within and between the different slide types are large. Generally, the rate of movement and total displacement of a slide depends on the change in post-failure behaviour and the wetness of the material. Most materials initially experience little deformation and move as a few distinct, but interacting units at first. With increasing displacement, these units may break-up and the material disintegrates. If the material is not restrained in its movement, high speeds can be attained on steep slopes. Debris slides transform into debris avalanches in this manner or into debris flows when the material is wet and liquefies. Both types can move at high speeds and cover great distances. Equally, fine-grained material can transform into a mudslide if the material is sufficiently wet. Although the velocity of mudslides is typically much lower than those of debris avalanches or flows, debris flows are erratic events that affect steep slopes whereas mudslides remain active in one area over a much longer period and at significantly lower slope angles.

The down-wear of sliding rock is less extreme as in the case of the free-fall movement in topples and slides. Notwithstanding, the compression of water or air in the pore space may lead to fluidisation, which reduces the available frictional strength, or results in the loss of the intrinsically high rock strength. High speeds of down-wear have been reported (Hutchinson and Bandhari 1971).

Translational slides usually travel larger distances than rotational slides because the latter can reach a new equilibrium by rotation of the unstable soil mass. When destabilised material empties on the lower slopes, it can move more freely and eventual run-out distances are controlled by the velocity of the destabilised material, the angle and resistance along its track and the material strength. In the case of liquefied cohesive materials, drainage is another important control. If pore pressures cannot dissipate, the material will remain in a liquid state and the run-out distance will be greater.

Slides are often episodic but may be so for different reasons: many shallow slides move seasonally due to increased pore pressures and elevated moisture contents after the wet season (Figure 3.9). Many larger slides that consist of several interacting units, such as multiple and successive slides, exhibit intricate spatio-temporal deformation patterns. Unloading at the base, for example due to undercutting, may reactivate the upslope part of a landslide and destabilise sections of the crown. In turn, the reactivated units will transfer their loads downslope and displace material at the toe that can be subsequently eroded.

Figure 3-9. Shallow rotational slide affected by changes in pore pressures on an embankment of the M25, near London, UK (Photo: J.E. Norris).

Recognition

Crozier (1973) defined seven morphometric indices for mass movements and found that rotational slides were distinguished from other mass movements by: the classification index (ratio of the true depth of the landslide compared to the overall length) and the tenuity index (the ratio of the length of displaced material to the concave part of the scarp and flank). Crozier's (1973) values for the classification index agreed with those of Skempton (1953), ranging from 0.15 to 0.27 for slopes between 13° and 28°. Based on these findings, a lower limit of 0.1 is commonly taken to distinguish rotational slides from translational ones (Selby 1993).

The location of rotational slides can often be inferred from detailed topographic maps by the presence of irregular, wavy contour lines and the concave shape of the scarp is shown by curved, closely spaced contour lines (Rib and Liang 1978). In the field, concave scarps in freshly exposed material, reversed slopes with water ponding behind them, generally disturbed and immature drainage patterns and the deviant orientation of soil and rock layers with respect to the stable part of the slopes are clear indicators (Crozier 1984; Table 3-8; Figure 3.10).

Translational slides are often arcuate, triangular or square in shape (Table 3-9). Their slip planes are long compared to their depth and movement takes place parallel to the slope. Scarps and flanks are often stripped from vegetation and soil, exposing the bedrock or parent material. The main body and

Table 3-8. Diagnostic features of rotational slides (after Rib and Liang 1978; Crozier 1984; Buma and Van Asch 1996).

Potential	**Relict**
1. Slopes that are moderately steep and of sufficient height to allow rotational movement; 2. Disturbance of the slope by undercutting or surcharges; 3. Uniform deposits of cohesive materials or severely broken down rock; 4. The presence of a weaker basal layer, particularly for the formation of multiple rotational slides; 5. A rise in pore pressures due to undrained loading, changes in the water regime (e.g., leaking sewerage) and rainfall; 6. Dynamic loading (seismic events, vibrations due to heavy traffic etc.).	1. Hummocky terrain; 2. Deposition area can be identified as raised ground with a steep front where the toe is located; 3. Reversed or gently sloping ground is found at the crown and the scarp is recognisable by barren soil or bedrock; 4. Drainage patterns on the affected slope and in the deposition area may show signs of disturbance; 5. Ponding and deposition of washed debris and organic material is found in depressions; 6. Tension cracks may be observed at the head.

Figure 3-10. Large slump or rotational slide (flat surface directly under scar in the level area with two sheds) in lacustrine deposits, induced by river undercutting; Voralberg, Austria (Photo: K. Smit Sibinga-Lokker).

deposition area of larger slides are often broken down into several interacting units separated by vertical escarpments or cracks. The toe buries the original surface and may be rolled over in a lobate shape. Drainage on translational slides is generally less disturbed than on rotational slides but streams or gullies tend to descend along the flanks and incise the slid material into the slip plane.

Slopes susceptible to sliding are moderately steep. On short slopes of sufficient height, the unstable soil mass is forced to rotate. On longer, straight slopes, the material moves more or less parallel to the surface, resulting in translational slides. The presence of softer or less permeable layers favours the occurrence of both rotational and translational slides. Abrupt changes in the topography and convergence increase respectively the triggering effect of seismic events and excessive rainfall or snowmelt.

Consequences and mitigation

True slides are by far the most common type of mass movements on natural and man-made slopes. Their consequences vary because of the difference in size and post-failure behaviour: damage to structures can be severe if a slide travels large distances, develops high speeds e.g., due to liquefaction, or experiences considerable differential deformation. Slides can be extremely dangerous when they catch people unawares. This is most

Table 3-9. Diagnostic features of translational slides (based partly on Rib and Liang 1978; Crozier 1984).

Potential	**Relict**
1. Slopes that are moderately steep and are of sufficient length to allow translational movement; 2. Slopes that are straight or slightly convergent in plan or exhibit a clear break of slope; 3. The presence of soil layers of varying or decreasing strength or permeability *or* the presence of multiple discontinuities in bedrock; 4. Disturbance of the slope by under-cutting or surcharges; 5. A rise in pore pressures due to undrained loading, changes in the water regime e.g., leaking sewerage, and rainfall; 6. Dynamic loading (seismic events, vibrations due to heavy traffic etc.).	1. Hummocky or stepped terrain with cracks that tend to follow the contour lines; 2. The scarp and flanks are near vertical near the crown and become more planar and gently sloping in the lower part. They expose barren soil or bedrock that are slowly recolonised by vegetation; 3. The landslide body is composed of several units of soil or rock that tend to become more frag-mented downslope; 4. Deposition area can be identified as raised ground with a lobate front where the toe is located; 5. Deposition area consists of material that has been rolled over or flows over the topography, burying the surface topography.

likely when a slide is rare e.g., rock slides (Sorriso-Valvo and Gullà 1996) and signs of progressive failure such as cracks opening at the future crown and bulging are not heeded. The mitigation against large landslides requires extensive and costly countermeasures but small, frequent landslides may equally inflict substantial damage over larger areas (Veder 1981; Schuster 1996).

On natural slopes, soil slips affect many places, especially where vegetation has been removed by forest fire or logging, or where the slope and drainage have been changed due to construction of access roads. Areas at risk should be identified by terrain reconnaissance and care should be taken to prevent or mitigate against such landslides if these areas cannot be avoided.

Due to the placement of fill and/or the construction of short, steep cuts, man-made slopes are extremely vulnerable to rotational slides. Such changes are also capable of reactivating pre-existing slides that have long been dormant and are not easily recognised (Chandler et al. 1973).

2.3.3 Flows

Description

Flows are viscous deformations of slope material in which all particles move at different rates and velocities decrease with depth. The material can behave as a Newtonian or non-Newtonian fluid: in the former case it is incapable of sustaining any shear stresses whereas in the latter case, the norm for flows, viscous deformation only occurs when a yield stress is exceeded (visco-plastic or *Bingham material*; Carson 1971).

Flows can be found in any type of slope material but rock flow or *sagging* is extremely slow and can be considered as a type of *creep* (Bisci et al. 1996). Flow distinguishes itself from creep by having discrete boundaries or narrow peripheral zones experiencing shear. Moreover, flows move at velocities that are a manifold of those of creep, which is generally imperceptible except to observations of long duration (Summerfield 1991). Creep occurs in response to the shear stress induced by overburden or is the net downslope transport of material as the result of episodic heave and settlement produced by solution, freeze-thaw, warming and cooling and wetting and drying cycles. Creep can also be caused by the biological activity of plant roots and grazing or burrowing animals loosening surface material (Selby 1993). Solifluction is a process similar to creep in which saturated material flows along extremely gentle slopes (>1°). Creep can rearrange particles and reduce the available resistance between them. Creep is therefore often a precursor to landsliding with the material accelerating until failure occurs (Summerfield 1991).

Flows are often the result of other mass movements e.g., landslides, falls or topples, where the material breaks up and pore pressures increase. The most common types are debris flows and soil flows (mudflows; Figure 3.11). These flows comprise different materials and require substantial amounts of water for fluidisation. However, soil flows can also occur in dry sands as a particular form of fluidisation (cohesionless grain flow or sand run). Such flows are very rare but can be potentially destructive due to their speed (Summerfield 1991; Schrott et al. 1996).

Debris flows (Figure 3.12) are composed of coarse material (gravel and boulders) which is embedded in a finer matrix (sand, silt, clay) with varying quantities of water that move as a slurry downslope (Corominas et al. 1996). Wet soil flows resemble debris flows except that they are composed of a single, fine grain-size (Schrott et al. 1996).

Figure 3-11. Mudflow in alpine meadow after heavy rainfall at Voralberg, Austria (Photo: L.W.S. de Graaff).

Flows originate from a source area where enough water is present to fluidise the available material. Fluidisation can occur after the debris covering the source area is mobilised as a slide (Hutchinson 1988), or when runoff laden with fines infiltrates, lifts and entrains the accumulated coarser material in the source area (Corominas et al. 1996; Blijenberg 1998). The fluidised material moves along the main track and is usually confined to the existing drainage pattern (Selby 1993). Rare, large and extremely rapid flows may have sufficient momentum to cross watersheds, e.g., after the break-through of landslide dammed lakes or following volcanic eruptions (*lahars*).

Figure 3-12. Talus slopes, fed by rolling and falling rock from steep cliffs, incised with debris flow channels with debris levees and debris flow fans (Pastoruri valley, Cordillera Blanca, Peru) (Photo: E. Cammeraat).

Along the track, some coarse material is pushed towards the side of a debris flow to form levees. Equally, debris flow material may be pushed out during the event, leading to fining upward sequences and clast-supported beds when the matrix is washed out (Corominas et al. 1996). Due to buoyancy, some boulders may concentrate on top of the deposits (Bagnold 1954). If no differentiation occurs, debris flow deposits have a chaotic appearance with the clasts floating in the matrix (Johnson and Rodine 1984). Deposition occurs where the gradient becomes sufficiently low and where the flow material is no longer confined, debris fans may develop. Wet soil flows exhibit the same morphology but due to their more uniform composition sedimentary differences in their deposits are not easily observed. Both wet soil flows and debris flows are the intermediaries between non-liquefied slides and hyper-concentrated stream flow (Pierson and Costa 1987). In debris flows and wet soil flows, the thickness of the shear zone increases compared to slides and viscous behaviour dominates but, in contrast to stream flows, the central zone still tends to move as a rigid plug and water is not a transporting medium.

Snow avalanches are another type of flow. Although they are mainly composed of snow and ice they move more or less similar to flows in geologic slope materials and may include or entrain a substantial part of the latter.

Causes

For the sustained activity of debris and wet soil flows, a continuous source of material is needed in combination with steep slopes. Such a supply may be found in (formerly) glaciated areas in the form of moraines and proglacial deposits, at the lower limit of alpine discontinuous permafrost, in soil mantled couloirs or weathering pockets, or underneath steep cliffs or on talus slopes (Schrott et al. 1996). Depressions or hollows are preferred sites for the initiation of flows because of the accumulation of material and the convergence of streamlines, which leads to elevated pore pressures. Since many source areas lie above the tree line, vegetation if present, provides little root reinforcement in deep colluvial soils (Dietrich et al. 1986).

To become wet flows, the materials need to be reworked and incorporate excessive amounts of water that can be delivered by intensive rainfall, rapid snowmelt and more rarely lake or glacier overflows (Selby 1993). Rainfall intensity and duration determine largely the initiation of many landslides and relationships describing the threshold of debris flow occurrence in terms of rainfall intensity, duration and frequency have been defined with and without consideration of the antecedent moisture conditions in different environments (Caine 1980; Sidle et al. 1985; Blijenberg 1998).

Movement

The activity of flows is controlled by the rate of accumulation of material in the source area and the frequency of potential triggers (Corominas et al. 1996). Upon triggering, the head collapses with rapid flow along the track and the deposition of material at the accumulation lobe. During movement, undrained loading within the flow mass leads to constant changes in the velocity of the mass. Flows can be extremely rapid: high velocities can be reached and values in excess of 10 ms^{-1} are common (Johnson and Rodine 1984; Costa 1984; Hutchinson 1988). Because of their momentum, flow tracks can extend over many kilometers, even at low gradients. Debris flows can erode their channel and thus increase their volume significantly (Jibson 1989). Deposition only occurs when the gradient decreases and excessive pore pressures dissipate. The deposition threshold and the final thickness of the lobe are determined by the cohesion of the visco-plastic flow (Johnson and Rodine 1984).

Recognition

Characteristics of wet flows and those of sites that are potentially subject to these mass movements are given in Table 3-10.

Consequences and mitigation

Debris flows are common phenomena in high mountain environments where they can incur substantial damage to infrastructure and threaten lives. During intense episodes of debris flow activity, they may choke river systems and increase the risk of sudden surges of hyper-concentrated flows (torrents). The consequences of such events and lahars can be catastrophic.

Wet soil flows are often subordinate features of other mass movements. However, their consequences can be serious due to the large displacement involved. Essential to the mitigation against flows is the recognition of possible source areas, the likely track ways and the probable extent of the deposition areas. Countermeasures against flows may include the construction of check dams and grids along the track or the regulated evacuation of material over the debris fan. Wet soil flows can additionally be controlled by the drainage of potential source areas.

Table 3-10. Diagnostic features of wet flows (after Rib and Liang 1978; Costa 1984; Corominas et al. 1996; Schortt et al. 1996).

Potential	Relict
1. Steep slopes; 2. Availability of loose debris and/or fines; 3. Poor drainage as evidenced by high drainage density, impervious substrate or infiltration impeded by permafrost; 4. Absence or sparse vegetation cover; 5. Intense rainfall or rapid snowmelt; 6. Flooding, irrigation or fluctuations in reservoir levels; 7. Volcanic eruptions; 8. Possibility of earthquakes or vibrations.	1. Scarp is typically funnel-shaped or serrated. Upper part is long and narrow and bare and striated when fresh. The crown may show few cracks; 2. The track is sinuous, long and narrow and follows the existing drainage patterns; 3. Infilling is evident: coarse material in finer matrix (debris flow) or conical heap of soil (soil flow); 4. Levees may be present in the middle and lower part of the track; 5. At the toe, material spreads in lobes. Debris flows may have a steep front if material was relatively dry.

3. EROSION

3.1 Introduction

Soil erosion, mainly due to water, is a growing problem that affects all European countries. Water erosion affects approximately 115 million hectares, which constitutes 12% of the European surface, and about a fifth has also been eroded by wind (42 m ha^{-1}). The effects of erosion are translated into a direct reduction of soil productivity and into a significant degradation of the ecosystem's dynamics and functions. With a very slow rate of soil formation, it has been calculated that any loss of more than 1 ton $ha^{-1}yr^{-1}$ can be considered irreversible in 50-100 years time (Van Lynden 1994).

Erosion is a natural process and is a geological phenomenon that can be accelerated by humans due to adverse land use techniques. It is therefore important to consider the rate at which soil erosion occurs, especially in comparison to weathering and soil infiltration rates. It is clear that in many places, soil erosion rates are higher than the weathering rates and that over time a large amount of soil will be lost.

The Mediterranean region is one of the areas that suffers from this process of accelerated erosion, reaching at places to irreversible levels of degradation. The loss of the fertile topsoil by erosion leads to a deterioration of soil quality. This has an important effect on the biomass production, which will lead to loss of crop production. Irreversible land degradation at a human time scale, resulting in loss of soil productivity in dryer climates, is also known as desertification (Brandt and Thornes 1996), and especially the southern part of Mediterranean Europe is threatened by this process. Desertification is one of the major environmental threats for dryland regions all over the world suffering from soil erosion, desiccation and salinization. In the Mediterranean region, water erosion can result in soil losses of up 20-40 ton ha^{-1} in individual storms, and with losses of more than 100 ton ha^{-1} in extreme events (EEA 1999).

Other seriously affected areas are Northern and Eastern Europe and Northern China. More detailed information on research on erosion can be found in the textbook of Morgan (2005).

3.2 General principles

Soil erosion could be defined as the removal of the soil surface particles by water, wind, ice, or other geological agents, including processes such as gravitational creep. Erosion is a natural and continuous process. Soils are created through weathering processes where geomorphic surface mechanisms are insignificant in relation to the rate of soil formation. When soil surface processes become more important, weathered parent materials and soils will be removed and transported, and the material will be deposited elsewhere.

Generally, soils with faster infiltration rates, higher levels of organic matter and good structure, have a greater resistance to erosion. Sand, sandy loam and loam-textured soils tend to be less erodible than silt, very fine sand, and certain clay textured soils. The susceptibility of a soil to be eroded or affected by erosion has been defined as "soil erodibility" (Wischmeier et al. 1971).

The influence of human activities has favoured the development of erosion processes at a greater speed than normal, natural, geological erosion. This phenomenon is known as “accelerated erosion”.

The main parameter that promotes the development of accelerated erosion, in many cases, is the degradation or loss of the vegetation cover; this can be caused by forest fires, deforestation or, more immediately, as a result of overgrazing or construction activities. On the other hand, land abandonment and forest fires, particularly in marginal areas, intensify the effects of this process together with the use of still inappropriate agricultural practices. Loss of vegetation cover exposes soils to wind and water erosion, therefore loss of soils decreases soil fertility and the potential for vegetation production. The final result is a decrease in the carrying capacity of the land.

Vegetation cover is important for soil protection because:

- it reduces the kinetic energy of runoff and this favours water infiltration on soil
- plant roots hold the soil in position and protect it from being washed away
- it breaks the impact of raindrops, decreasing their erosive capacity (Andreu et al. 1998).

Although soil erosion affects most of the European landscape, its effects are especially important in areas that have a limited vegetation cover protection, such as in the drier parts of Europe and in agricultural areas where soils are kept uncovered between harvest and the initial growth phase of successive crops for the next growing season. Key strategies to combat erosion, such as afforestation, or legislation related to improved soil management practices, such as tillage, maintenance of hedgerows or the introduction of cover crops after the main crop is harvested, are still lacking in many areas of Europe.

3.3 Causes of soil erosion

Soil erosion by wind or, mainly, by water is a natural phenomenon that is in equilibrium with landscape and ecosystems dynamics. Human development acts on this equilibrium usually increasing the degree of the process.

The magnitude of soil erosion depends on:

- Climate, mainly due to rainfall characteristics (intensity, amount, etc). High intensity rainfalls in combination with scarce or absent vegetation cover increases the impact of erosion (erosivity).
- Initial soil moisture conditions of the topsoil. When topsoil is (nearly) saturated, overland flow may intensify the effect of erosion processes.
- Type of soil, whose physical and chemical characteristics determine its resistance to erosion (erodibility).

- Runoff, and its energy, is responsible for the removal and transport of soil particles previously detached by the destruction of soil aggregates caused by raindrop impact. If soils show lower infiltration capacity due to soil compaction, crusting or textural characteristics (silty or clayey soils), then runoff generation increases.
- Slope morphology, gradient and length. The steeper or longer the slope, the greater the energy of runoff and its capabilities of soil removal, increasing erosion potential.
- Human action, through changes in vegetation cover and agricultural practices (deforestation, inappropriate land preparation and management practices, etc) or direct action on the soil (compaction by heavy machinery, infrastructures, etc), are the major causes of soil degradation and increased erosion.
- Lack of crop rotation leading to loss of soil quality.
- Overgrazing and overstocking by animals can adversely affect the vegetation cover and increase soil erosion rates.
- Forest fires also reduce vegetation cover and removal of the burnt wooden stems leads to increased soil erosion (see Chapter 7).

Water erosion is also promoted by converting grassland to arable land and increasing field sizes by the removal of hedgerows. In both cases, previous obstacles to reduce runoff generation and its energy are destroyed allowing the free movement of water over the soil surface. Deforestation on steep slopes also affects erosion as the soil loses its protective cover from rain and runoff.

3.4 Processes of soil erosion

3.4.1 Sheet erosion

This form of erosion is characterised by the removal of a fairly uniform layer of soil from the land surface by runoff water or overland flow. The superficial soil horizon is removed from the slope in thin layers (sheets) and often disappears, gradually making it difficult to monitor because the damage is not immediately perceptible. This type of process could be considered as the initial step to developing other forms of erosion like rills, gullies or pipes. This process is very effective because it can cover large areas of sloping land and, if no other erosive forms appear, is often unobserved until the subsoil is exposed.

Sheet erosion is an important mechanism of slope degradation and source of sediment in cut slopes in granitic and andesitic soils. Highway cuts in these soils often give the impression of being stable e.g., no presence of

Figure 3-13. Exposed tree roots resulting from sheet erosion over slightly sloping crusted terrain (Korsimoro, Burkina Faso) (Photo: E. Cammeraat).

rills/gullies, yet discharge tonnes of soil into roadside ditches (Gray and Sotir 1996). Protection of underlying soil layers is very important because these layers contain the majority of soil nutrients, humus and other fertility components.

Sheet erosion (Figure 3.13) produces the loss of the finest soil particles which contain the majority of plant-available nutrients and organic matter, affecting the productivity of the land. It may also result in removal of seeds or seedlings and reduce the soil's ability to store water for plants to draw upon between rainfall events. Another characteristic of soils affected by this kind of erosion process is the appearance of soil crusts. Crusts are produced by the accumulation of fine particles derived from the break down of aggregates, into which air and water can no longer penetrate.

Soil deposited off-site through this type of erosion could cause crop and pasture damage, water-quality deterioration and stream, dam, lake and reservoir sedimentation. This soil deposition could be a sign of the incidence of sheet erosion together with the appearance of surface flow patterns (Figure 3.14), soil pedestals protected by the root mass of the plants and, in the last erosive stages, the presence of light-coloured subsoil appearing on the surface. Soils which are repeatedly cultivated, abandoned fields and fallow soils or soils that are bare through overgrazing by stock or pest animals are particularly vulnerable (Figure 3.15).

Figure 3-14. Appearance of surface flow patterns in a vineyard (Photo: V. Andreu).

Figure 3-15. Erosion on repeatedly cultivated soil (eroded vineyard) (Photo: V. Andreu).

3.4.2 Rill or gully erosion

Rill erosion is the removal of soil by water from very small but well-defined, visible channels or streamlets where there is a concentration of overland flow (Gray and Sotir 1996). In general, rill erosion is more serious than sheet erosion, and it is most accentuated when intense storms occur in watersheds or sites with high runoff-producing characteristics and loose, shallow topsoil. Rills are small enough to be easily removed by normal tillage and grading operations.

Rill erosion (Figure 3.16) often occurs with sheet erosion, and is the most common form of water erosion. It is often described as the intermediate stage between sheet and gully erosion, and occurs by a concentration of runoff or overland flow into deeper, faster-flowing channels, which follow depressions or low points through the soil. The shearing power of water flow can detach and remove soil particles starting the development of these channels, which can reach depths of 0.3 m. Once these structures are formed, they become the preferred routes for sediment transportation. Soil removed by runoff water from these streamlets runs through land with poor surface drainage, forming many smaller channels only a few centimetres deep. Rill erosion usually appears on recently cultivated soils, and can often be observed in between

Figure 3-16. Rill erosion in an almond orchard, Sierra de Torrecilla, SE Spain (Photo: E. Cammeraat).

crop rows. The effects of rill erosion can be easily removed by tillage, but it is a process most often overlooked until it becomes a major problem.

Rill erosion is commonly observed on agricultural land devoid of vegetation and so is often seen in paddocks, cropping areas after tillage, or recently cultivated soils following high-intensity rainfalls, which is the typical situation of traditional Mediterranean dry farming. After intense rains, cultivated topsoil overlying denser cohesive subsoil or compacted layers often exhibit rill erosion. Poorly managed pasture areas where overgrazing occurs, on texture-contrast (duplex) soils are also susceptible.

Gully erosion could be considered as an advanced stage of rill erosion, where surface channel gullies (intermittent stream channels larger than rills) have been eroded to the point where they cannot be smoothed over by normal tillage operations. In this process, runoff water is accumulated in narrow channels and, depending on the intensity of the rainfall, can gradually remove the soil from the channels increasing their depths, reaching from about 0.3 m to as much as 30 m. Gullies tend to form where large volumes of runoff are concentrated and discharged onto steep slopes with erodible soils e.g., undefended culvert outlets. Gully erosion is common in grasslands whilst in steep, forested watersheds, gullies are the main form of

Figure 3-17. Example of gully erosion in an abandoned olive grove (Carcavo catchment, SE Spain) (Photo: E. Cammeraat).

erosion (Gray and Sotir 1996). Gully formation is frequently characterised by steep sidewalls and a lack of vegetation. The maximum depth to which gullies are cut is governed by topography, by resistant layers in the soil, by bedrock, or by the local base level. Many gullies develop head wards; i.e., they extend up the slope as the gully deepens in the lower part.

3.4.3 Piping and tunnel erosion

Underground (groundwater) erosion is the removal of soil caused by groundwater seepage or movement towards a free face. It is also known as piping and occurs as a result of bank drainage or, in general, when seepage forces exceed intergranular stresses or cohesive forces (Gray and Sotir 1996). Pipes can form in the downstream side of earth dams, gully heads, streambanks, and slopes where water exits from the ground. Once a cavity (pipe) forms, it is able to enlarge quickly since the flow follows the path of low flow resistance.

This type of erosion process usually appears in soils with subsurface horizons that allow free water penetration and movement through more than the surface layers. It occurs in two main ways:

- Water infiltrates through a porous medium producing enough drag force to transport material at the outlet through liquefaction or Coulomb failure. It could favour the formation of a subsurface channel that works back from the outlet, often developing a complex branched network (Figures 3.17 and 3.18).
- Produced by a progressive expansion of an existing channel or macropore, which can include enlargement of animal burrows, root channels, desiccation or unloading cracks, occurs mainly due to the shear stress exerted by flowing water.

The first process is generally known as piping, properly, whereas the second one has been identified as tunnel erosion (Bryan and Jones 1997; Zhu et al. 2002). The main practical difference is that tunnel erosion characteristics do not necessarily develop from the channel, although sediment must be evacuated, and they do not necessarily involve high discharge pressures. Both phenomena are favoured by the presence of appreciable exchangeable sodium. However, both terms are used indistinctly (Dunne 1990; Piccarreta et al. 2006).

The consequence in the evolution of this process is, generally, that the disproportionate enlargement of the section of the channel or tunnel near the inlet may form a funnel-shaped feature that, reaching the limit of resistance of the geological materials, will collapse producing a gully or cleft of great proportions. This process usually appears:

Figure 3-18. Expanding gully system initiated by piping (Photo: E. Cammeraat).

- In areas characterised by steep slopes and an excess of water, which develop organic soils. In some circumstances, desiccation cracks could provide the pathways for piping initiation.
- In soils with degraded vegetation cover and compaction produced by livestock trampling. In this case, infiltration hampers localised overland flow. This is usually observed in degraded semi-arid rangelands.
- Zones dominated by sodic materials, mainly on smectites. Desiccation cracks are common and the resistance of subsoil materials to fluid shear stress is low. It corresponds to badland areas on arid and semi-arid environments.
- On bench-terraced soils with poor cohesive materials at the subsoil. The pipes develop at the edge of the bench terrace until they collapse, and afterwards destroy the retaining wall.

3.4.4 Tillage erosion

Tillage of land leads to movement of soil particles by the farmer. When tillage is carried out on slopes this leads to a net downward movement of soil particles. On the top of fields, soil is removed and is accumulated on the downslope sides (see Figure 3.19). The ploughing direction is also important (down-up hill or transverse along the hillside) (Takken et al. 2001).

This process acts at considerable rates and was neglected until the last ten years. Theories are currently being refined and made applicable in soil tillage management (Quine and Zhang 2004). On coarse textured soils, a sieving effect can also occur, where the coarsest particles are concentrated on the topsoil (Poesen et al. 1998).

Figure 3-19. Embankment at the upper slope of an agricultural field resulting from tillage erosion, in Northern Ecuador (Photo: J.P. Lesschen).

3.4.5 Wind erosion

This process is defined as the breakdown of solid rock into smaller particles and its removal by wind. It may occur on any soil whose surface is dry, unprotected by vegetation (to bind it at root level and shelter the surface) and consists of light particles. The mechanisms include straight-forward picking up of dust and soil particles by the airflow and dislodging or abrasion of surface material by the impact of particles already airborne (EEA 2005). Its intensity and effects on soil directly depends on soil surface stability and protection, so texture, organic matter content, moisture, relief and vegetation cover become key parameters. Other important factors that affect the process are wind velocity, surface roughness and length and morphology of the area.

Wind erosion is especially important on areas characterized by fine sandy and silty soils (loess, marls, etc), with poor aggregates structure and scarce vegetation cover. Its main effect results in a reduction of soil fertility and damages to seedlings and crops, mainly in young plants (Figure 3.20).

Wind erosion has a more important impact on agricultural lands but does not have a critical influence on processes that can affect slope stability, such as mass movements, landslides or water erosion, which are the main subject of this Chapter. For this reason, the effects of wind erosion are not considered further. The reader is referred to the following publications for more information:

- USDA-ARS Wind Erosion Research Unit. Bibliography on wind erosion. http://www.weru.ksu.edu/new_weru/publications/publications.shtml
- Warren, A. A Bibliography of Wind Erosion and Related Phenomena, http://www.geog.ucl.ac.uk/~awarren/wnero.pdf
- Thomas E. Gil T.E., Warren A., Stout J.E. Bibliography of Aeolian Research (1646-2007). http://www.lbk.ars.usda.gov/wewc/biblio/bar.htm
- Favis-Mortlock, D. June 2005. "The Soil Erosion Site" http://soilerosion.net/

Figure 3-20. Shallow sand dunes, resulting from local wind erosion processes, invading an olive orchard in E. Morocco (Photo: E. Cammeraat).

4. REFERENCES

Andreu V, Rubio JL, Cerni R (1998) Effects of Mediterranean shrub cover on water erosion. J Soil Water Conserv 53:112-120

Angeli MG, Gaspareto P, Menotti RM, Pasuto M, Soldati M (1996) Rock avalanche. In: Dikau R, Brunsden D, Schrott L, Ibsen M (eds) Landslide Recognition: Identification, Movement and Causes. Wiley, Chichester pp 190-201

Azzoni A, Barbera GL, Zaninetti A (1995) Analysis and prediction of rockfalls using a mathematical model. Int J Rock Mech Mining Sci 32:709-724

Bagnold RA (1954) Experiments on a gravity free dispersion of large solid spheres in a Newtonian fluid under shear. Royal Soc London Proc Series 225:49-63

Bisci C, Dramis, F, Sorriso-Valvo, M (1996) Rock flow. (Sackung). In: Dikau R, Brunsden D, Schrott L, Ibsen M (eds) Landslide Recognition: Identification, Movement and Causes. Wiley, Chichester pp 150-160

Blijenberg HM (1998) Rolling stones?: triggering and frequency of hillslope debris flows in the Bachelard Valley, southern French Alps. PhD thesis, Utrecht University, The Netherlands

Bogaard TA (2001) Analysis of hydrological processes in unstable clayey slopes. PhD thesis, Utrecht University, The Netherlands

Bozzolo D, Pamini R (1986) Simulation of rock falls down a valley side. Acta Mech 63: 113-130

Brandt CJ, Thornes, J (eds) (1996) Mediterranean Desertification and Land Use. Wiley, Chichester

Broilli L (1974) Ein Felssturz in Grossversuch. Rock Mech Suppl 3:69-78

Brunsden D (1984) Mudslides. In: Brunsden D, Prior DB (eds) Slope Instability. Wiley, Chichester pp 363-418

Buma J, Van Asch ThWJ (1996) Slide (rotational). In: Dikau R, Brunsden D, Schrott L, Ibsen M (eds) Landslide Recognition: Identification, Movement and Causes. Wiley, Chichester pp 43-61

Bryan RB, Jones JAA (1997) The significance of soil piping processes: inventory and prospect. Geomorphology 20:209-218

Caine N (1980) The rainfall intensity-duration control of shallow landslides and debris flows. Geografiska Annaler 62A:23-27

Campbell RH (1975) Soil slips, debris flows and rainstorms in the Santa Monica Mountains and vicinity, southern California. USGS Professional paper 851

Cannon SH, Kirkham RM, Parise M (2001) Wildfire-related debris-flow initiation processes, Storm King Mountain, Colorado, Geomorphology 39:171-188

Carson MA (1971) The Mechanics of Soil Erosion. Pion, London

Chandler RJ, Pachakis M, Mercer J, Wrightman J (1973). Four long-term failures of embankments founded on areas of landslip. Quarterly J Eng Geol 6:405-422

Chandler RJ (1986) Processes leading to landslides in clay slopes: a review. In: AD Abrahams (ed) Hillslope Processes. Allen & Unwin Inc, Boston pp 343-360

Clowes A, Comfort P (1982) Process and Landform. Conceptual Frameworks in Geography. Oliver & Boyd, Edinburgh

Cooke RU, Doornkamp JC (1990) Geomorphology and Environmental Management: a New Introduction. 2nd edn. Clarendon Press, Oxford

Coppin NJ, Richards IG (1990) Use of Vegetation in Civil Engineering. CIRIA, Butterworth, London

Corominas J (1996) Debris slide. In: Dikau R, Brunsden D, Schrott L, Ibsen M (eds) Landslide Recognition: Identification, Movement and Causes. Wiley, Chichester pp 97-102

Corominas J, Remondo R, Farias P, Estevao M, Zézere J , Días de Terán J, Dikau R, Schrott L, Moya J, González A (1996). Debris flows. In: Dikau R, Brunsden D, Schrott L, Ibsen M (eds) Landslide Recognition: Identification, Movement and Causes. Wiley, Chichester pp 161-180

Costa JE (1984) Physical geomorphology of debris flows. In: Costa JE, Fleisher PJ (eds) Developments and Applications of Geomorphology. Springer-Verlag, Berlin pp 268-317

Crozier MJ (1973) Techniques for the morphometric analysis of landslips. Zeitschrift für Geomorphologie NF, 17:78-101

Crozier MJ (1984) Field assessment of slope Instability. In: Brunsden D, Prior DB (eds) Slope Instability. Wiley, Chichester pp 103-142

Crozier MJ (1986) Landslides: Causes, Consequences and Environment. Croom Helm, London

Cruden DM, Varnes DJ (1996) Landslide type and processes. In: Turner AK, Schuster RL (eds) Landslides: Investigation and Mitigation. TRB. National Academy Press. Washington, DC pp 36-75

Dai FC, Lee CF (2001) Terrain-based mapping of landslide susceptibility using a geographical information system: a case study. Can Geotech J 38:911-923

Day RW (1997) Case studies of rockfall in soft vs. hard rock. Env Engin Geosc 3:133-140

Dietrich WE, Wilson CJ, Reneau SL (1986) Hollows, colluvium and landslides in soil-mantled landscapes. In: Abrahams A (ed.) Hillslope Processes, Allen & Unwin, Winchester pp 361-388

Dijkstra TA, Van Asch ThWJ, Rappange FE, Meng X (2000) Loess slope stability modelling. In: Derbyshire E, Meng XM and Dijkstra TA (eds) Landslides in the Thick Loess Terrain of North-West China. Wiley, Chichester pp 203-219

Dikau R, Brunsden D, Schrott L, Ibsen M-L (1996a) Landslide Recognition: Identification, Movement and Causes. Wiley, Chichester

Dikau R, Brunsden D, Schrott L, Ibsen M-L (1996b) Introduction. In: Dikau R, Brunsden D, Schrott L, Ibsen M-L (eds) Landslide Recognition: Identification, Movement and Causes. Wiley, Chichester pp 1-12

Dikau R, Schrott L, Dehn M (1996c) Topple. In: Dikau R, Brunsden D, Schrott L, Ibsen M-L (eds) Landslide Recognition: Identification, Movement and Causes. Wiley, Chichester, pp 29-42

Dorren LKA (2003) Mountain geoecosystems: GIS modelling of rockfall and protection forest structure. PhD thesis, University of Amsterdam, The Netherlands

Dorren LKA, Maier B, Putters US, Seijmonsbergen AC (2004) Combining field and modelling techniques to assess rockfall dynamics on a protection forest hillslope in the European Alps. Geomorphology 57:151-167

Dorren LKA, Berger F, Le Hir C, Mermin E, Tardif P (2005) Mechanisms, effects and management implications of rockfall in forests. For Ecol Manag 215:183-195

Douglas GR (1980) Magnitude-frequency study of rockfall in Co. Antrim, N. Ireland. Earth Surf Proc 5:127-129

Dunne T (1990) Hydrology, mechanics, and geomorphic implications of erosion by subsurface flow. In: Higgins CG, Coates DR (eds) Groundwater geomorphology, the role of subsurface water in earth-surface processes and landforms. Geol Soc Am Spec Pap 252:1-28

EPOCH (1993) Temporal Occurrence and Forecasting of Landslides in the European Community. Flageollet J-C (ed). 3 Volumes

Erismann TH (1986) Flowing, rolling, bouncing, sliding, synopsis of basic mechanisms. Acta Mech 64:101-110

Erismann TH, Abele G (2001) Dynamics of Rockslides and Rockfalls. Springer-Verlag, Berlin

EEA-European Environment Agency (1999) 3.6. Soil Degradation. In: Environment in the European Union at the turn of the century. State of Environment report No 1/1999. EEA, Copenhagen

EEA-European Environment Agency (2005) SOER 2005 – The European environment – State and outlook 2005. EEA, Copenhagen

Evans SG, Hungr O (1993) The assessment of rockfall hazard at the base of talus slopes. Can Geotech J 30:620-636

Flageollet JC, Weber D (1996) Fall. In: Dikau R, Brunsden D, Schrott L, Ibsen M (eds) Landslide Recognition: Identification, Movement and Causes. Wiley, Chichester pp 13-28

Goodman RE (1980) Rock Mechanics. Wiley, New York

Gostelow TP (1996) Landslides. In: Singh VP (ed) Hydrology of Disasters. Kluwer Publishers, Rotterdam pp 183-230

Gray DH, Sotir RB (1996) Biotechnical and Soil Bioengineering Slope Stabilization – A Practical Guide for Erosion Control.Wiley, New York

Guthrie RH (2002) The effects of logging on frequency and distribution of landslides in three watersheds on Vancouver Island, British Columbia. Geomorphology 43:273-292

Hearn G, Barret RK, McMullen ML (1992) CDOT Flexpost rockfall fence development, testing, and analysis. Rockfall prediction and control and landslide case histories. Transport Research Record 1343. National Academy Press, Washington, DC pp 23-29

Hétu B, Gray JT (2000) Effects of environmental change on scree slope development throughout the postglacial period in the Chic-Choc Mountains in the northern Gaspé Peninsula, Québec. Geomorphology 32:335-355

Hungr O, Evans SG (1988) Engineering evaluation of fragmental rockfall hazards. In: Bonnard Ch, Landslides/Glissements De Terrain, Proceedings of the 5th International Symposium on Landslides in Lausanne. Balkema, Rotterdam, pp 685-690

Hutchinson JN (1988) Morphological and geotechnical parameters of landslides in relation to geology and hydrology, General report. In: Bonnard, Ch (ed) Landslides/Glissements De Terrain, Proceedings of the 5th International Symposium on Landslides in Lausanne. Balkema, Rotterdam pp 3-35

Hutchinson JN, Bhandari RK (1971) Undrained loading, a fundamental mechanism of mudflows and other mass movements. Geotechnique 21:353-358

Ibsen M-L, Brunsden D, Bromhead E, Collison, A (1996a) Block slide. In: Dikau R, Brunsden D, Schrott L, Ibsen M (eds) Landslide Recognition: Identification, Movement and Causes. Wiley, Chichester pp 64-78

Ibsen M-L, Brunsden D, Bromhead E, Collison, A (1996b) Slab slide. In: Dikau R, Brunsden D, Schrott L, Ibsen M (eds) Landslide Recognition: Identification, Movement and Causes. Wiley, Chichester pp 79-85

Jahn J (1988) Entwaldung und Steinschlag. In: Proceedings of the International Congress Interpraevent Graz 1988, Bnd 1:185-198

Jibson RW (1989) Debris flows in Southern Puerto Rico. Geol Soc Am Spec Paper 236:29-56

Johnson AM, Rodine JR (1984) Debris flows. In: Brunsden D, Prior DB (eds) Slope Instability. Wiley, Chichester pp 257-362

Jomelli V, Francou B (2000) Comparing the characteristics of rockfall talus and snow avalanche landforms in an Alpine environment using a new methodological approach: Massif des Ecrins, French Alps. Geomorphology 35:181-192

Kienholz H, Mani P (1994) Assessment of geomorphic hazards and priorities for forest management on the Rigi north face, Switzerland. Mountain Res Dev 14:321-328

Kirkby MJ, Statham I (1975) Surface stone movement and scree formation. J Geol 83: 349-362

Lambe TW, Whitman RV (1979) Soil Mechanics, SI version. Wiley & Sons, New York

Morgan RPC (2005) Soil Erosion and Conservation, 3rd edn. Blackwell, Malden

Nieuwenhuis JD (1991) Variations in stability and displacements of a shallow seasonal landslide in varved clays. PhD thesis, University Utrecht, The Netherlands Balkema, Rotterdam

O'Loughlin CL (1974) The effect of timber removal on the stability of forest soils. J Hyd NZ 13:121-123

Ortiz JMR, Serra J, Oteo C (1986) Curso aplicado de cimetaciones. 3rd Edition, Colegio Oficial de Arquitectos de Madrid

Patton FD (1970) Significant factors in rock slope stability. In: Van Rensburg PWJ (ed) Planning open pit mines. South African Institute of Mining and Metallurgy. Johannesburg pp 143-151

Peila D, Pelizza A, Sassudelli F (1998) Evaluation of behaviour of rockfall restraining nets by full scale tests. Rock Mech Rock Eng 31:1-24

Piccarreta M, Faulkner H, Bentivenga M, Capolongo D (2006) The influence of physico-chemical material properties on erosion processes in the badlands of Basilicata, Southern Italy. Geomorphology 81:235-251

Pierson TC, Costa JE (1987) A rheological classification of subaerial sediment-water flows. In: Costa JE, Wieczorek GF (eds) Debris flows/Avalanches: Process, Recognition and Mitigation. Geol Soc Am, Rev Eng Geol, VII: pp 1-12

Poesen JW, van Wesemael B, Bunte K, Solé Benet A (1998) Variation of rock fragment cover and size along semiarid hillslopes: a case-study from southeast Spain. Geomorphology 23:323-336

Quine TA, Zhang Y (2004) Re-defining tillage erosion: quantifying intensity–direction relationships for complex terrain. 2. Revised mouldboard erosion model. Soil Use Manag 20:124-132

Rib HT, Liang T (1978) Recognition and identification. In: Schuster RL, Krizek RJ (eds) Landslides analysis and control. Transport Research Board, Special Report 176, Academy Press, Washington, DC pp 34-80

Ritchie AM (1963) Evaluation of rockfall and its control. Highway Research Record 17, Highway Research Board, National Research Council, Washington, DC pp 13-28

Rupke J, Cammeraat E, Seijmonsbergen AC, van Westen CJ (1988) Engineering geomorphology of the Widentobel catchment, Appenzell and Sankt Gallen, Switzerland. A geomorphological inventory system applied to geotechnical appraisal of slope stability. Eng Geol 26:33-68

Schrott L, Dikau R, Brunsden D (1996) Soil flows. In: Dikau R, Brunsden D, Schrott L, Ibsen M (eds) Landslide Recognition: Identification, Movement and Causes. Wiley, Chichester pp 181-188

Schumm SA, Chorley RJ (1964) The fall of the threatening rock. Am J Soil Sci 262: 1041-1054

Schuster RL (1996) Socioeconomic significance of landslides. In: Turner AK, Schuster RL (eds) Landslides: Investigation and Mitigation. TRB. National Academy Press. Washington D.C pp 12-31

Seijmonsbergen AC (1992) Geomorphological evolution of an alpine area and its application to geotechnical and natural hazard appraisal – in the NW. Ratikon mountains and S. Walgau (Vorarlberg, Austria). PhD thesis, University of Amsterdam, The Netherlands

Selby MJ (1993) Hillslope Material and Processes. 2nd Edition. Oxford University Press, Oxford

Sidle RC, Pearce AJ, O'Loughlin CL (1985) Hillslope stability and land use. American Geophysical Union. Water Resources Monograph 11

Sidle RC, Dhakal AS (2002) Potential effects of environmental change on landslide hazards in forest environments. In: Sidle RC (ed) Environmental Change and Geomorphic Hazards in Forests, IUFRO Research Series, No. 9, CAB International Press, Oxen, UK pp 123-165

Skempton AW (1953) Soil mechanics in relation to geology. Proc Yorks Geol Soc 29: 33-62

Skempton AW (1964) The long-term stability of clay slopes. Géotechnique 14:95-102

Soriso-Valvo M, Gullà G (1996) Rock slides. In: Dikau R, Brunsden D, Schrott L, Ibsen M (eds) Landslide Recognition: Identification, Movement and Causes. Wiley, Chichester pp 85-96

Statham I (1976) A scree slope rockfall model. Earth Surf Proc 1:43-62

Statham I, Francis SC (1986) Influence of scree accumulation and weathering on the development of steep mountain slopes. In: Abrahams AD (ed) Hillslope Processes, Allen and Unwin, Winchester pp 245-267

Summerfield MA (1991) Global Geomorphology: An Introduction to the Study of Landforms. Longman, Harlow

Takken I, Govers G, Jetten V, Nachtergaele J, Steegen A, Poesen, J (2001) Effects of tillage on runoff and erosion patterns. Soil Till Res 61:55-60

Terzaghi K (1962) Stability of slopes on hard unweathered rock. Géotechnique 12:251-270

Van Asch ThWJ, Van Steijn H (1991) Temporal patterns of mass movements in the French Alps. Catena 18:515-527

Van Beek LPH, Van Asch ThWJ (2004) Regional assessment of the effects of land use change on landslide hazard by means of physically based modelling. Nat Hazards 31: 289-304

Van Beek LPH, Wint J, Cammeraat LH, Edwards JP (2005) Observation and simulation of root reinforcement on abandoned Mediterranean slopes. Plant Soil 278:5-74

Van Lynden GWJ (1994) The European Soil Resource: Current status of soil degradation causes, impacts and need for action. Council of Europe. Strasbourg

Varnes DJ (1978) Slope movements: type and processes. In: Schuster RL, Krizek RJ (eds) Landslides Analysis and Control. Transport Research Board, Special Report 176, Academy Press, Washington, DC pp 11-33

Varnes DJ (1984) Landslide hazard zonation: a review of principles and practice. Commission on Landslides, IAEG 3, UNESCO, Paris

Veder C (1981) Landslides and their Stabilisation. Springer Verlag, Wien

Whalley WB (1996) Rockfalls. In: Dikau R, Brunsden D, Schrott L, Ibsen M (eds) Landslide Recognition: Identification, Movement and Causes. Wiley, Chichester pp 217-256

Wieczorek GF (1996) Landslides triggering mechanisms. In: Turner AK, Schuster RL (eds) Landslides: Investigation and Mitigation. TRB. National Academy Press. Washington, DC pp 76-90

Wischmeier WH, Johnson CB, Cross BV (1971) A soil erodibility nomograph for farmland and construction sites. J Soil Water Conserv 26:189-192

Ziemer RR, Swanston DN (1977) Root strength changes after logging in South East Alaska. Pacific Northwest Forest and Range Experiment Station Research Note, PNW-306. Forest Service, USDA, Portland, 10 pp

Zhu TX, Luk SH, Cai QG (2002) Tunnel erosion and sediment production in the hilly loess region, North China. J Hydrol 257:78-90

Zinggeler A, Krummenacher B, Kienholz, H (1991) Steinschlagsimulation in Gebirgswäldern. Berichte und Forschungen 3: Geographisches Institut der Universität Freiburg pp 61-70

Chapter 4

HOW VEGETATION REINFORCES SOIL ON SLOPES

Alexia Stokes[1], Joanne E. Norris[2,3], L.P.H. van Beek[4], Thom Bogaard[5], Erik Cammeraat[6], Slobodan B. Mickovski[7], Anthony Jenner[8], Antonino Di Iorio[9], Thierry Fourcaud[10]

[1] INRA, AMAP, A A-51/PS2, Boulevard de la Lironde, 34398 Montpellier cedex 5, France, [2] Halcrow Group Limited, Endeavour House, Forder Way, Cygnet Park, Hampton, Peterborough, PE7 8GX, U.K., [3] School of Architecture, Design and Built Environment, Nottingham Trent University, Burton Street, Nottingham, NG1 4BU, U.K., [4] Department Physical Geography, Faculty of Geosciences, Utrecht University, P.O. Box 80115, 3508TC Utrecht, The Netherlands, [5] Water Resources Section, Department Civil Engineering and Geoscience, Delft University of Technology, P.O. Box 5048, 2600 GA Delft, The Netherlands, [6] IBED-Physical Geography, University of Amsterdam, Nieuwe Achtergracht 166, 1018 WV Amsterdam, The Netherlands, [7] Jacobs UK Ltd., Glasgow G2 7HX, U.K., [8] Milton Keynes Parks Trust, Campbell Park Pavilion, 1300 Silbury Boulevard, Campbell Park, Milton Keynes, MK9 4AD, U.K., [9] Dept. Chemical and Environmental Sciences, University of Insubria, Via Valleggio 11, CO-22100, Italy, [10] CIRAD, AMAP, A A-51/PS2, Boulevard de la Lironde, 34398 Montpellier cedex 5, France

Abstract: *Once the instability process e.g. erosion or landslides has been identified on a slope, the type of vegetation to best reinforce the soil can then be determined. Plants improve slope stability through changes in mechanical and hydrological properties of the root-soil matrix. The architecture of a plants root system will influence strongly these reinforcing properties. We explain how root morphology and biomechanics changes between species. An overview of vegetation effects on slope hydrology is given, along with an update on the use of models to predict the influence of vegetation on mechanical and hydrological properties of soil on slopes. In conclusion, the optimal root system types for improving slope stability are suggested.*

Key words: root architecture, root reinforcement, anchorage, tensile strength, erosion, landslides, slope hydrology, model

J.E. Norris et al. (eds.), Slope stability and erosion control: Ecotechnological solutions, 65–118.

1. INTRODUCTION

Vegetation can act as a protective barrier between the soil and the natural elements which stimulate erosion or mass movement. Plants exhibit many different forms and structures, but in general the elements that are likely to be useful in ecotechnological solutions to slope stability are:

1. Roots, to provide anchorage and absorb water and nutrients from the soil.
2. Stems, to support the above-ground parts and capture eroding soil.
3. Leaves, to intercept precipitation and initiate evapotranspiration leading to decreased soil moisture levels (Coppin and Richards 1990).

Certain types of plants are intrinsically better suited than others for specific stabilization objectives. Table 4-1 gives desirable characteristics for the "ideal" functions of vegetation. It is unlikely that the "ideal" species will be available for the exact needs of a slope stability problem. Each species will produce a different rooting pattern and different amounts of above ground biomass depending on site conditions and climate. In this chapter we will discuss how vegetation can be used to stabilize and fix soil on slopes. Although riverbank stability is an extremely important area to consider, a fairly large body of literature exists on this subject, so will not be treated here (e.g. Schiechtl and Stern 1996; Abernethy and Rutherfurd 2001; Simon and Collison 2002).

Table 4-1. Desirable plant characteristics for functions of vegetation (after Gray and Sotir 1996).

Function	Desirable Plant Characteristics
Capture and restrain	Strong, multiple, and flexible stems; rapid stem growth; ability to re-sprout after damage; ready propagation from cuttings and root suckers
Cover and armour	Extensive, tight, and low canopy; dense, spreading, surface growth; fibrous root mat
Reinforce and support	Multiple, strong, deep roots; rapid root development; high root/shoot biomass ratio; good leaf transpiration potential
Improve habitat	Shade and cover to moderate temperatures and improve moisture retention; soil humus development from litter; nitrogen fixation potential

1.1 Types of vegetation

<u>Grasses</u>

Grasses are very quick growing and offer a dense protective ground cover. Due to their meristem being at ground level, moderate damage to the plant does not cause lasting damage and fast regrowth can occur. Grasses with their dense network of shallow roots are usually useful in protecting sites from surficial erosion (Gray and Sotir 1996). However, some species have very deep root systems e.g. vetiver (*Vetivaria zizanoides* L.) and are renowned for their suitability in the restoration of unstable and eroded slopes (http://www.vetiver.org).

Herbs

Herbs (herbaceous plants) have little or no woody tissue. Herbs can be annuals or perennials and in the latter case, lose their leaves in winter. They tend to grow closer to the ground providing a dense ground cover with a shallow root system.

Woody plants and shrubs

A woody plant has a perennial woody stem and supports vegetative growth. Many annuals appear to form woody stems in their first year, but nonetheless die back. Shrubs are defined as low-growing woody plants with multiple stems. Shrubs can vary in height depending on species from 0.2 m to up to 6.0 m. In areas where visibility is essential shrubs could be preferred to trees as they will not grow as large and be easier to control and maintain. Although root systems may not spread as deep and as far as tree root systems, tensile strength may be comparable, depending on the species (Table 4-4).

Trees

Trees are perennial woody plants having a main stem and usually a distinct crown. Depending on soil type, tree roots can grow up to several metres deep and wide (Stone and Kalisz 1991). Therefore, trees are often considered suitable for reinforcing soil on slopes. However, if soil is shallow, tall trees are more susceptible to falling over during wind storms, thus reducing slope stability.

Vegetation responds in different ways to different environments. Growth on slopes may be difficult, especially in mountainous regions where resources may be limited and extreme weather events common. Combined with abiotic stresses, growth conditions can be harsh. Nevertheless, some plant species are well adapted and in this chapter we will discuss how plants acclimatize to life on slopes, with an emphasis on root growth. A large body of literature already exists concerning plant response to the climatic conditions encountered in mountains (see e.g. Körner 2003) and will not be discussed here.

1.2 Plant response to abiotic stress

Abiotic stress can be defined as an external, non-biological load imposed on a plant which may result in a modification of growth processes. These changes in plant growth can improve stem stiffness (Telewski 1995) or root anchorage on a slope (Chiatante et al. 2003). Several abiotic stresses exist which may have an influence on tree and plant growth on slopes. These forces include wind loading, erosion, mass movement of soil, avalanches, debris flow and runoff. The way in which a tree or a plant responds will have consequences for the subsequent growth and anchorage on the slope (Table 4-2).

When a woody plant or tree is subjected to an abiotic stress, a corresponding strain results (Telewski 1995). Two types of strain, *elastic* and *plastic*, may be manifested in different parts of the stem, branches and roots of the structure. Elastic strain represents a reversible change, after which the structure returns to its original state. Tree stem displacement due to e.g. wind loading, where the stem returns to vertical following the event, is an example of elastic strain. In plastic strain, the change which occurs is irreversible, and results in damage to the tree or woody plant. Permanent stem displacement or rupture during a storm event, or after a landslide or avalanche, is an example of this type of strain.

Although the growth adaptation of plants and trees to abiotic stress is an accepted scientific phenomenon (Telewski 1995), details of the way in which it occurs are still not clear (Telewski 2006). The term used to describe the acclimative growth response of plants and trees to mechanical loading was named 'thigmomorphogenesis' by Jaffe (1973). 'Thigmo' from the Greek 'to touch' and 'morphogenesis' implying the changes incurred during growth. The first experiments carried out by Jaffe (1973), Jaffe et al. (1980), and Jaffe and Telewski (1984) investigated the effects of touching, brushing, rubbing and flexing herbaceous species. Although not exactly realistic, these mechanical perturbations can be likened to dynamic loading e.g. wind loading or frequent soil mass movement on a slope. Typical responses included an increase in stem taper, a reduction in branch length and changes in wood anatomy. The increase in stem taper is usually achieved by a reduction in stem elongation and/or an increase in radial growth (Telewski 1995). The resulting plant may therefore have a "stunted" appearance, thus decreasing the speed-specific drag of the crown. The first studies combining the effects of wind action on root growth were carried out on Sitka spruce (*Picea sitchensis* Bong Carr.) and European larch (*Larix decidua* Mill.) by Stokes et al. (1995, 1997). Results showed that changes in root system morphology and topology increased anchorage in young trees subjected to wind loading. Roots held in tension during loading were more numerous and branched than those held in compression, which can become thicker and more rigid (Stokes 1999). Extra secondary thickening and anatomical changes may also occur in zones of high mechanical stress, which reduces the likelihood of failure (Nicoll and Ray 1996; Stokes and Guitard 1997; Di Iorio et al. 2007). Trees and woody plants growing on slopes are in a similar loading situation and changes in root system architecture have also been found to occur (Chiatante et al. 2003; Section 1.4.2).

In response to static loading, e.g. a slow build-up of snow or debris behind a tree growing on a slope, a tree can form reaction wood which serves to right the tree if leaning, or if the centre of gravity is offset (Figure 4.1; Timell 1986). Reaction wood may be formed in the stem, branches and sometimes the roots of woody plants and trees (Patel 1964; Timell 1986; Hsu et al. 2006). In angiosperms, this wood forms in the mechanically stressed zones held in tension and is called tension wood, whereas in gymnosperms, compression wood is found in the zones held in compression. Both types of wood are anatomically, chemically and physically different to normal wood, and have huge consequences for the technological quality of the timber. Reaction wood formation is often accompanied by the laying down of new wood in the most mechanically stressed areas of the structure, resulting in an eccentric cross-section (Figure 4.1) which will also increase stiffness along the axis of bending (Telewski 1995).

Few correlations between external abiotic stress and root response have been identified for trees growing on sloping sites. A study by Scheichtl (1980) suggests that roots growing uphill are stronger than their counterparts downhill due to differences in tissue structure. A series of experiments conducted by Shrestha et al. (2000) concluded that lateral roots elongate uphill on sloping sites, with increasing slope angle leading to increased uphill growth, which has also been observed in mature Downy oak (*Quercus pubescens* Willd.) (Di Iorio et al. 2005). However, studies by Khuder et al. (2006) on Black locust (*Robinia pseudoacacia* L.) seedlings inclined at different angles showed that little root growth occurs uphill. Nicoll et al. (2006) studying mature Sitka spruce even showed that root growth was preferential across the slope, but suggested that abiotic forces e.g. wind loading, are more likely to influence root architecture than slope angle.

Table 4-2. The abiotic forces to which vegetation is subjected on a slope, along with the induced acclimative response and consequences for mechanical stability.

Process	***Stress***	***Plant Response***	***Consequences***
Wind forces: Prevailing (i.e. unidirectional and sustained	Static	Increased stem taper and changes in anatomy Changes in root architecture and anatomy	Resistance to breakage Modified root anchorage characteristics
	Dynamic	Reduced crown surface area	Decreased drag coefficient
Frequent gusting (high turbulence)		As above and stem damping reaction	Elastic strain, allowing a return to equilibrium state following event
Mass movements: Landslide (short timescale)	Static/ Dynamic	Tension/compression forces in roots	Modified root anchorage characteristics
Landslide (long timescale)	Static	Tension/compression forces in roots	Modified root anchorage characteristics, leads to soil reinforcement Formation of reaction wood, strengthening roots
Rockfall	Dynamic	Stem damping reaction Scar formation in broadleaf species	Buttress formation, leading to arching Elastic strain, allowing a return to equilibrium state following event Resistance to pathogens
Surcharge changes: On vegetation (affects branch weight, e.g. snowfall)	Static	Increased stem and branch bulk, at high strain nodes Annual leaf loss (in some species)	Increased stem and branch strength, hence resistance to plastic strain Reduced area for weight accumulation
On ground (affects stem e.g. debris accumulation)	Static	Stem buttress formation Changes in root architecture	Increased stem strength Resistance to overturning
Runoff	Static/ Dynamic	Surface root disturbances	Reduced root reinforcement of soil in localised areas Reduced root anchorage strength
Erosion processes	Static	Surface root disturbances Drying-out of roots	Reduced root reinforcement of soil in localised areas Reduced root anchorage strength
Avalanches	Static/ Dynamic	Leaning stem Tension/compression forces in roots Stem damping reaction	Formation of reaction wood to right the stem Modified root anatomy Elastic strain, allowing a return to equilibrium state following event

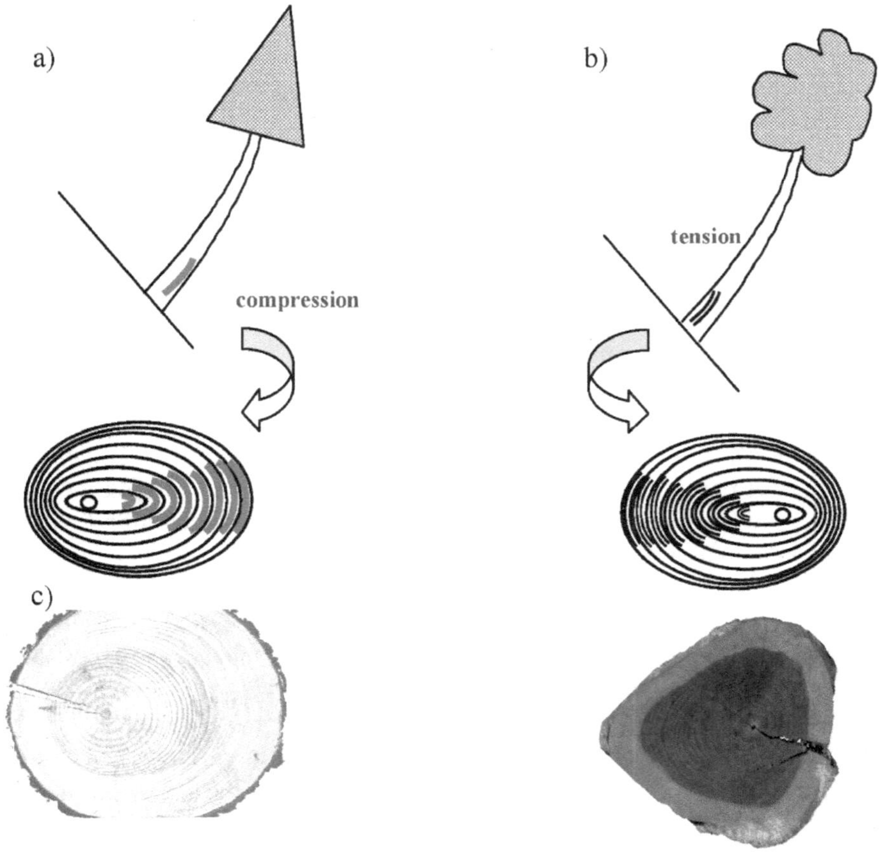

Figure 4-1. Reaction wood forms in the mechanically stressed zones of stems and branches, which have been permanently displaced due to e.g. wind or snow loading. In a) conifers, this wood forms in the zones held in compression and in b) broadleaf species, the zones held in tension. Reaction wood formation is usually accompanied by extra secondary growth, resulting in c) an eccentric cross-section (Photos: A.D. Kokutse).

1.3 Hydrological factors influencing root reinforcement

1.3.1 Introduction

Vegetation has an important influence on hillslope hydrology, and vice-versa, thus influencing the activity of erosion and landslides:

- Canopy reduces the amount and the intensity of rainfall reaching the soil by interception;

- Vegetation depletes the soil moisture storage by transpiration;
- Vegetation cover and litter on the soil protects the soil surface and prevents the formation of crusts, thus maintaining the infiltration capacity and enhancing evaporation from the litter layer;
- Vegetation improves the soil structure by rooting and litter production and promotes soil biological activity. This results in meso- and macropores that augment the infiltration capacity.

As an example, Jetten (1994) calculated that the removal of trees in a tropical forest increased the percolation to the groundwater from 50% to 80% while the evapotranspiration decreased by 30%. Combined, these effects reduced the amount of water that the soil has to accommodate, thus lowering erosion activity. They also improved soil infiltration and increased its storage capacity. As a consequence, water in the topsoil can be transported faster to greater depths. Following a rainfall event this can shorten the time during which soil moisture conditions favour shallow landslides significantly, but it can equally lead to faster groundwater recharge. The subsequent rise in pore pressure may trigger landslides at greater depths.

Therefore, strong bonds exist between vegetation and hydrological behaviour. Changes in hydrological patterns e.g. changes in soil moisture content within a slope, can result in modifications in vegetation patterns (e.g. Ridolfi et al. 2003) or even in the internal structure of individual woody plants (Barij et al. 2007). Vegetation itself also creates environments where water is trapped and stored in the soil, especially in semi-arid and sub-humid environments where competition for water is important (Valentin et al. 1999; Rietkerk et al. 2004) e.g. bamboo forests have been cited as having significantly increased soil moisture and air humidity, thus improving local environmental conditions (Storey 2002; Stokes et al. 2007b). Regional climate may even be influenced in some cases (e.g. Dekker et al. 2007).

1.3.2 The hydrological process

The hydrological cycle

The hydrological system is a closed water balance system driven by solar energy. The salt water ocean is the final and largest store of water. The fresh water cycle is generally on a shorter time base (except the deepest groundwater systems), and consists of clouds, snow, lakes, soil and groundwater. On a smaller spatial scale of a watershed or hillslope the water balance is:

$$P = Q + ET + \Delta S \qquad (1)$$

where P is precipitation, Q is discharge, ET is evapotranspiration and ΔS is the changes in water storage. The evapotranspiration term includes most of the influence of vegetation, such as interception, evaporation of the intercepted water, transpiration of soil water consumed by the roots of vegetation, etc. The main pathways of water on a hillslope are indicated in Figure 4.2 (see Kirkby 1978).

Precipitation

Precipitation, as measured by rain gauges, is called total precipitation or gross precipitation. Precipitation includes rainfall, snow, hail and sleet, and is therefore a more general term than rainfall, which is only the liquid state. Several hydrological processes like interception, surface storage and infiltration make sure that not all gross rainfall is discharged.

Net precipitation is the amount of precipitation reaching the ground under a vegetative cover, thus, gross precipitation minus interception loss, corrected for stemflow. Effective precipitation is used in agriculture and is defined as that part of the total precipitation falling on an irrigated area that is effective in meeting the consumptive use requirements i.e. available for crops. Rainfall excess is the volume of rainfall available for direct runoff and is equal to the total rainfall minus interception, depression storage, and absorption. In hydrology the latter definition is more often used reversed.

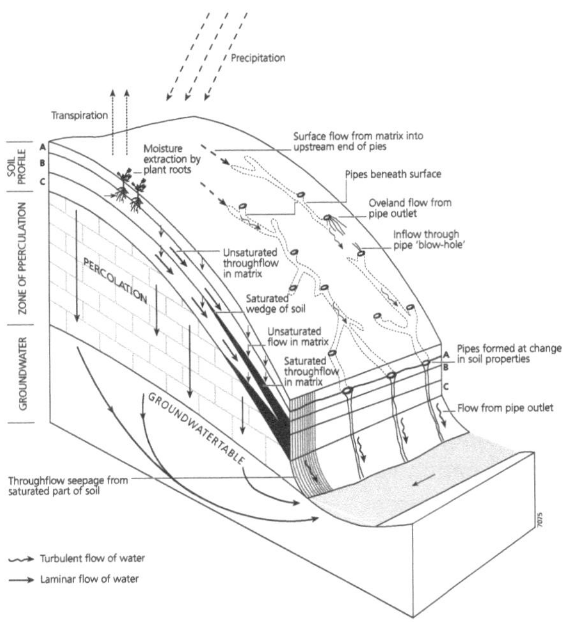

Figure 4-2. Routes of subsurface flow on a hillslope.

Interception

Interception diminishes the rainfall that is available for infiltration (net rainfall). Any water that is captured on the plant may evaporate and not be available for infiltration. This fraction is known as the gross interception (Zinke 1967).

Interception losses depend primarily on the ability of plants to detain rainfall. Water that is not intercepted by the vegetation is passed to the surface (free throughfall). As the rain continues the storage capacity of the vegetation is exceeded and drainage will occur as water drips from the leaves (dripfall) and runs along branches and stems (stemflow). Because both the time to saturation and the drainage processes are related to the effective rainfall intensity, most interception equations have the general appearance of a curvilinear relationship that is bounded by the storage capacity of the vegetation.

Interception of rainfall occurs at all vegetation levels. Rainfall not lost to interception at the canopy level may be intercepted by the undergrowth and litter that cover the soil. Compared to canopy interception the latter quantity is more difficult to measure in the field and is often accommodated by inclusion with the actual evapotranspiration. However, recently Gerrits et al. (2006) measured beech litter interception and evaporation using a lysimeter approach. At the different levels interception may vary independently over time (e.g. in the case of deciduous forests).

Total throughfall can be measured directly by collecting the rainfall that passes through the vegetation canopy. In this case it includes some dripfall and the collected fractions of rainfall will be variable in space and time. Likewise, stemflow can be measured by collecting all water flowing along branches or stems. Alternatively, the fraction of free throughfall can be estimated from the leaf area index (LAI), the ratio of the leaf surface over the projected canopy area (LAI, $m^2 \cdot m^{-2}$). This method has the advantage that LAI is readily measured at the stand level from radiation measurements below and above the canopy (LI-COR 1992). Also, radiation measurements are quicker and more amenable than the physical collection of rain. Its drawbacks are that it neglects the influence of rainfall intensity and evaporation rates on the total available net rainfall.

Measurements of interception losses provide an indication of the likely rainfall losses. Available data concerns mainly tall or woody vegetation. Precious little information is available on the losses under herbaceous plants or turf. These values represent long-term averages and as a consequence may under- or overestimate the interception loss due to the natural variability in rainfall intensity and due to temporal and spatial variability in vegetation conditions (open stands).

Stemflow and dripfall concentrate water at the base of stems or under the canopy and may lead to erosion problems due to splash and overland flow.

Also, they wash down the products from atmospheric deposition and plant material that may be either beneficial or adverse to vegetation health (Likens et al. 1977).

Infiltration and runoff generation

The rate by which water can infiltrate into the soil is composed of a constant infiltration capacity that is linked to intrinsic soil properties and a variable contribution related to the matric suction or *sorptivity* of the topsoil (Parlange and Smith 1976). Initially, the matric suction dominates the infiltration rate during a rainfall event (suction controlled infiltration) but as more water enters the soil its influence is less felt and the rate decreases asymptotically to the constant rate of the infiltration capacity (gravity controlled infiltration). Runoff will occur when the available net precipitation exceeds the infiltration rate (infiltration excess – or *Hortonian* overland flow; Horton 1933, 1945). This runoff leads to overland flow that can infiltrate again or cause erosion. Any rainfall excess and entrained sediment that are not stored along the slope will be discharged to the channel.

The high precipitation rates needed for Hortonian overland flow are generally met by the high rainfall intensities in Mediterranean areas and the tropics or after rapid snowmelt. In highly permeable soils, rainfall excess and Hortonian overland flow are rare. In those areas, runoff occurs when the storage capacity of the soil is exceeded. This may happen locally (saturation excess overland flow) or result from saturated lateral throughflow (return flow). Saturated lateral throughflow requires that the vertical drainage is impeded in which case water is transported rapidly downslope through the more permeable topsoil. Short-lived episodes of saturated lateral flow in immediate response to rainfall have been observed in the permeable topsoils of forested hillslopes, especially in temperate regions (subsurface storm flow). This rapid redistribution of moisture along the slope is extremely important for the generation of positive pore pressure at potential slip planes and as a constituent of peak channel flow. A direct expansion of the concepts of subsurface stormflow and return flow is the theory of contributing areas (e.g. Hewlett and Hibbert 1967). As rainfall increases, a saturated zone will develop in the soil which accumulates and progresses upward. This explains the toe failures of many slopes. The theory of contributing areas fails, however, to explain the observation of saturated areas higher on a slope. Here, existing groundwater bodies can be enlarged or perched groundwater bodies generated that may lead to failure. Betson (1964) described the theory of partial areas indicating that small areas (e.g. 5-10%) within a catchment produce more than 50% of the runoff. These areas, not necessarily in the valley bottom, are very important also when dealing with hydrological triggering of landslides.

1.3.3 Concepts of matric flow and preferential flow

The problems of erosion and flooding have instigated much research on infiltration and runoff in hillslope hydrology. Less attention has been paid to percolation and groundwater recharge. Most process studies of unsaturated zone hydrology have been undertaken from an agricultural viewpoint and consider only the topsoil albeit in detail. Groundwater recharge was given less attention. Consequently, a gap exists in the process knowledge between infiltration, percolation and groundwater behaviour at the hillslope scale.

Agricultural hydrological research has improved our understanding of the behaviour of water in the unsaturated zone. The water transport in the unsaturated zone has for a long time been described similar to saturated groundwater flow, i.e. as continuous flow domain through the matric pore space. This is described by the Darcy-Buckingham equation:

$$q = k(h)\nabla H \qquad (2)$$

where k(h) is the unsaturated hydraulic conductivity as a function of the matric potential h. With increasing h (decreasing moisture content), k(h) generally diminishes, and, ∇ H is the gradient of the total potential in the *x*, y and z direction.

According to this concept, new water 'pushes' old water downwards (piston flow). Most numerical unsaturated models are based on the Richards equation for matric flow, which is an extension of the Darcy-Buckingham concept.

Many soils have a heterogeneous pore space and therefore matric flow concepts have been extended by macropore flow (see Beven and Germann 1982 for an overview of macropore flow). Macropores are areas within the soil where atmospheric pressure exists. Examples are tension cracks, fissures, dessication cracks, root holes, animal burrows, soil pipes, etc. The combination of matric porosity and macropore porosity is called the double porosity concept. In the macropores water flow behaves as open channel flow whereas in the matric Darcian flow prevails and the interaction between the two systems is highly complex.

The piston flow concept for matric flow is nowadays almost totally replaced by the concept of preferential flow. The words 'preferential flow' do not specifically refer to macropore or fissure flow, but more to preferred flow as a consequence of heterogeneity or state-dependent anisotropy, that is: prolonged wet (moist) 'subsurface fingers' transport water from the surface to the ground water system (wetting front instability). Preferential flow paths can develop as a result of (i) an increase of the soil hydraulic conductivity with depth, (ii) water repellency, (iii) redistribution of infiltration after the end of a rain shower or irrigation, (iv) air entrapment, (v) non-ponding rainfall (De Rooij 2000).

The consequence of preferential flow concept is that fluxes of water, nutrients and contaminants do not travel homogeneously but are concentrated along several flow paths with relatively high velocities. This results in faster transport than assumed under the piston flow assumption. For landslides the main consequence is that infiltrated water can reach the slip surface much faster than expected with Darcian flow conditions (van Beek and Cammeraat 2007).

1.3.4 Evapotranspiration and soil moisture conditions

Evapotranspiration occurs as heat at the soil surface is used to vaporise moisture. This moisture is lost as evaporation from the soil surface and as transpiration through vegetation. The eventual rate of evapotranspiration depends on the turbulence, a result of the wind distribution and surface roughness, that allows this water vapour to dissipate into the air. Evapotranspiration is therefore not constant over time but varies strongly with the atmospheric boundary conditions and the state of the soil surface and vegetation which influence the rates of evaporation and transpiration.

Potential evapotranspiration usually refers to the maximum amount of water that can be evaporated under the present atmospheric conditions from a uniform soil or water surface when the water supply is not a limiting factor (Doorenbos and Pruitt 1977; Brutsaert 1982). A well-known physically–based model that calculates the potential evapotranspiration under assumed boundary conditions from generally available atmospheric or climatic data is that by Penman (1948). Alternatively, reference potential evapotranspiration can be calculated from simpler functions that relate potential evapotranspiration to temperature and radiation (e.g. Makkink 1957; Priestly and Taylor 1972) or deduced from water balance calculations under controlled conditions, for example with lysimeters, or evaporation pans.

One of the influences of vegetation on evapotranspiration is a change in surface roughness. Especially in the case of isolated trees increasing turbulence leads to higher evapotranspiration rates. Also, when the water supply is limited, evapotranspiration may exhaust the available moisture. Plants will try to retain moisture by closing the stomata of their leaves. This increases the resistance against the transpiration and the actual evapotranspiration will be lower than the potential evapotranspiration. This concept provides the basis of the physically-based Penman-Monteith Equation which introduces an additional crop resistance in the water vapour exchange through turbulence.

Transpiration rates can be deduced from sapflow measurements. However, such measurements are only feasible for larger plants and trees and generally sparse which makes it difficult to capture the spatial and temporal heterogeneity in the vegetation cover in the crop resistance parameter. Therefore, the relationship between the actual evapotranspiration under a vegetation cover and the potential evaporation is mostly represented by a simple empirical constant, the crop factor, k_c (Doorenbos and Pruitt, 1977; Allen et al., 1998):

$$ET_C = k_c \cdot ET_0 \tag{3}$$

where ET_0 is the reference potential evapotranspiration [L·T-1], k_c is an empirical crop factor [-], and ET_C is the actual evapotranspiration by the vegetation.

The crop factor approach was developed originally for agriculture but it can be expanded to natural vegetation. It includes all vegetation effects on the evapotranspiration that arise through the characteristics of the individual plant or the plant community, including those of ground cover and surface roughness. The actual evapotranspiration comprises not only the transpiration but also the evaporation from the bare soil. This simplification is warranted as transpiration generally exceeds evaporation from a dry soil surface (Hooghart and Lablans 1988).

Crop factors are mostly not constant in time. They are a function of growth stage, soil moisture availability and vegetation health. From agricultural research detailed information on crop factors is available (Allen et al. 1998). Criticism about the method focuses on the simplified representation of the actual evapotranspiration as a constant fraction of the potential rate. It does not take the soil moisture availability explicitly into account. If soil moisture is highly variable and has a strong influence on the transpiration by plants, root water uptake can be described separately as a function of soil moisture (e.g. Feddes et al. 1978). Although this approach is coarse and simplistic, it is often in balance with the available data.

Lysimeters can be used to derive the crop factors but they often fail to contain representative samples of the vegetation. Consequently, their results are highly variable. An encouraging development in this respect is that with advances in remotely sensed data, high resolution estimates of the actual evapotranspiration over larger areas are available (SEBAL method, Bastiaanssen et al. 1998; Bastiaanssen 2000).

1.3.5 Volumetric changes of soils

Some soils, especially those with a high clay content, are vulnerable to swelling upon wetting and shrinkage upon drying. The processes of swelling and shrinkage can be reversed but may show some hysteresis. This is not the case in peaty soils where dessication is irreversible and leads to increased oxidation of the organic matter in the soil. Some common clay mineral types are more vulnerable to these volumetric changes than others. Especially montmorillonite clays and to a lesser degree illite clays are sensitive to this behaviour, whereas kaolinite is far less sensitive. When monovalent metal ions are present at the exchange complex of the clays, swelling is more important, especially in the case of the presence of sodium ions. The sensitivity to swelling can be directly translated to the dispersion behaviour of soils, which is an important aspect in soil crusting and soil erodibility. The physico-chemical background of these processes can be found in many textbooks such as Marshall and Holmes (1988).

Volumetric change of the soils can lead to irregular surfaces upon repeated wetting and drying, causing problems with regard to constructions e.g. highway embankments. When drying, the soil will be penetrated by deep open vertical cracks, which can be up to 20 cm wide and attain depths of over 1 metre. These cracks develop in the dry season as a reaction to soil moisture depletion by physical evaporation and due transpiration by plants, close, at least at the surface in the wet season. Cracks often reappear in the same places, as vertical crack surfaces are often covered with dust or silty sediments (Cammeraat 2002).

Cracks can be important preferential flowpaths of water. At the end of the dry season, a large rainfall event can cause water to be transmitted through these cracks towards the deeper solum, without saturating the whole soil. This water may accumulate deeper in the profile at the boundaries between the soil and the regolith or unweathered bedrock. If a perched watertable is developed on such a strong drop in vertical hydraulic conductivity, this may affect the stability of hillslopes. Other conditions such as mechanical properties and slope gradient are also in favour of this process. In this specific case soil water depletion by plants, causing deep shrinkage cracks, may be negatively affecting slope stability.

1.4 Mechanical factors

1.4.1 Introduction

The mechanical properties of vegetation have both adverse and beneficial effects on soil fixation and erosion. A balance of these effects must be maintained to ensure long-term soil stabilization. The role of vegetation in reinforcing and anchoring the soil contributes to its stability but is dependent on factors such as root system morphology, root strength, distribution, and root-soil interaction (Reubens et al. 2007).

1.4.2 Root system morphology

Root system morphology is complex and exhibits high variation, depending on species, soil type and site conditions (Coutts 1983a). Soil and site conditions which may affect morphology include:

- Availability of air and nutrients in soil;
- Soil moisture content and permeability;
- Location and variation of the groundwater table;
- Extent to which soil is compacted; and
- Presence of certain compounds in the soil (e.g. toxic substances, salinity).

When not limited by soil or ground water conditions, herbaceous, shrub and woody species have intrinsic root system morphological characters. Trees have been classified as having three main root system types: plate, heart and tap (Köstler et al. 1968; Figure 4.3). Plate root systems have large lateral roots and vertical sinker roots, heart systems possess many horizontal, oblique and vertical roots and tap systems one large central root and smaller lateral roots (see Chapter 6 for species list). Some species may be classed as having a mixture of root system types (Stokes 2002). In both broadleaved (Lyford 1980) and conifer (Preisig et al. 1979; Gruber 1994) tree species, the architecture of the root system, depending on soil conditions, can be modified from a tap rooted type to sinker and even very superficial root systems. Trees possessing heart and tap root systems have been classified as being the most resistant to uprooting and plate systems the least resistant (Stokes 2002; Dupuy et al. 2005a).

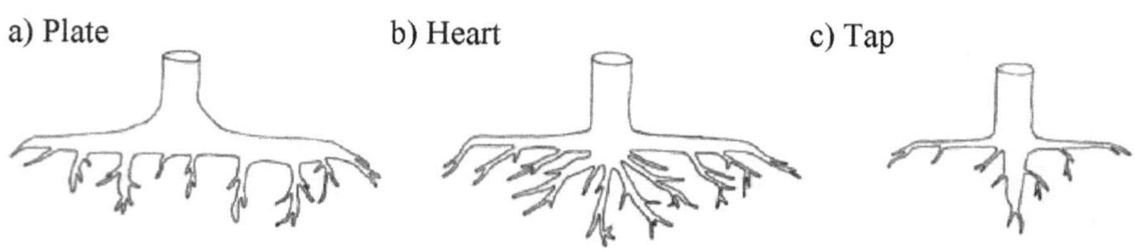

Figure 4-3. Different types of root system architecture a) 'plate' or 'sinker' system with large lateral roots and some smaller vertical roots, b) 'heart' system with many horizontal and vertical roots and c) 'tap' root system with one major central root and smaller horizontal and vertical roots (after Stokes and Mattheck 1996, reprinted by permission of the publisher).

When influenced by local soil conditions, e.g. the presence of a hard pan or a seasonal water table, rooting depth may be inhibited, and sinker or tap roots may be asphyxiated or unable to penetrate the hard pan (Nicoll and Ray 1996; Cucchi et al. 2004: Danjon et al. 2005). These root systems will thus have the appearance of a plate root system (Figure 4.4).

Figure 4-4. Reconstruction from 3D digitising data of a 50 year old tap rooted Maritime pine (*Pinus pinaster* Ait.) root system with vertical root growth impeded due to seasonal waterlogging and a layer of impenetrable hard pan (Image courtesy of F. Danjon/F. Lagane; see Danjon et al. 2005 for method).

Individual roots within a system may be further classified into subgroups depending on their morphology and function. Extensive roots are those which grow to large depths and spread diameters, while intensive roots are short, fine roots, localised within an area and often attached to larger structural roots. The term 'adventitious' refers to those lateral roots which originate from a woody parental root and grow at the soil surface; their specific function is the procuring of water and nutrients for the plant.

Root architecture is an important consideration in terms of the way in which forces on the tree structure are transferred into the ground. The shape of the root system ultimately determines the way in which these forces are distributed, be they dynamic or static (Coutts 1983a). The stability and soil holding capacity of trees on horizontal and sloping sites is strongly influenced by the symmetry of the structural system of woody roots. Three types of root system asymmetry exist:

- Type 1, whereby individual roots can vary in diameter, which can result in an asymmetric system, even if the arrangement of roots is regular (Figure 4.5a) (Coutts et al. 1999)
- Type 2, whereby the roots are not uniformly arranged, even though they may all be the same size (Figure 4.5b) (Coutts et al. 1999)
- Type 3, asymmetry (often found when growing on slopes), with irregular arrangement and variation of diameter (Figure 4.5c).

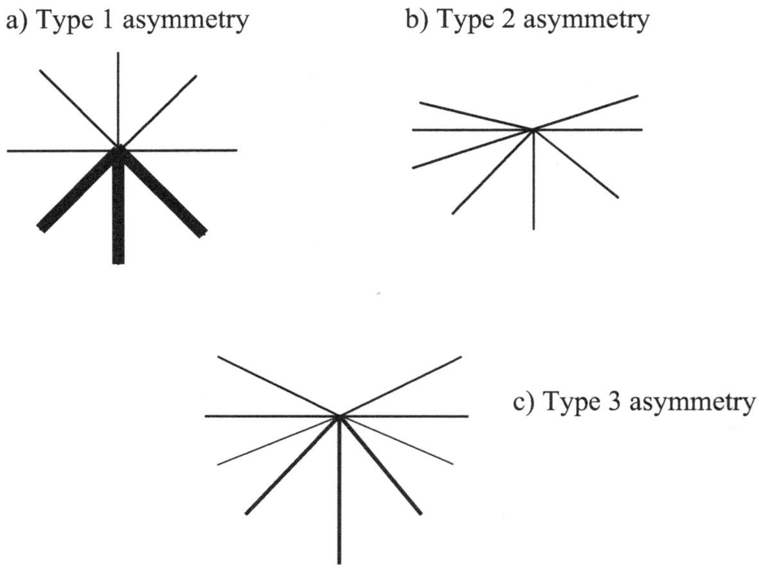

Figure 4-5. Root systems may exhibit different types of asymmetry: a) in Type 1, individual roots can vary in diameter, even if the arrangement of roots is regular; b) Type 2, whereby the roots are not uniformly arranged, even though they may all be the same size (modified from Coutts et al. 1999) and c) Type 3 on a slope, the arrangement of roots is irregular and roots vary in diameter. All tree root systems exhibit a combination of these asymmetries.

Tree stability is usually enhanced if root systems are symmetrical. However, trees on slopes tend to have highly asymmetrical systems, depending on species type (Nicoll et al. 2006). Trees can also respond to mechanical stress e.g. wind loading, by developing asymmetric root systems, with more numerous or thicker roots along the axis of the stress (Stokes et al. 1995; Mickovski and Ennos 2003). These trees will be better anchored, as long as the direction of the mechanical stress does not change. However, conflicting evidence exists concerning the asymmetric shape of root systems growing on slopes. Intuitively, it would be thought that root growth would increase on the up- and downhill sides of root systems, as roots in tension (uphill) are stronger than in compression. Roots on the downslope (compression) side of the tree could therefore be expected to be thicker in order to resist rupture during loading. But studies in the field on mature *P. sitchensis* have shown that root mass was concentrated across-slope on a 30° slope (Nicoll et al. 2006). Marler and Discekici (1997) found however that around 70% of roots of papaya (*Canica papaya* L.) on a 30° slope, formed on the downhill side. Watson et al. (1995) showed that in Kanuka (*Kunzia ericoides* (A. Rich)) and radiata pine (*Pinus radiata* D. Don), lateral roots were predominant up- and across-slope. In an elfin forest in Ecuador, Soethe et al. (2006) found

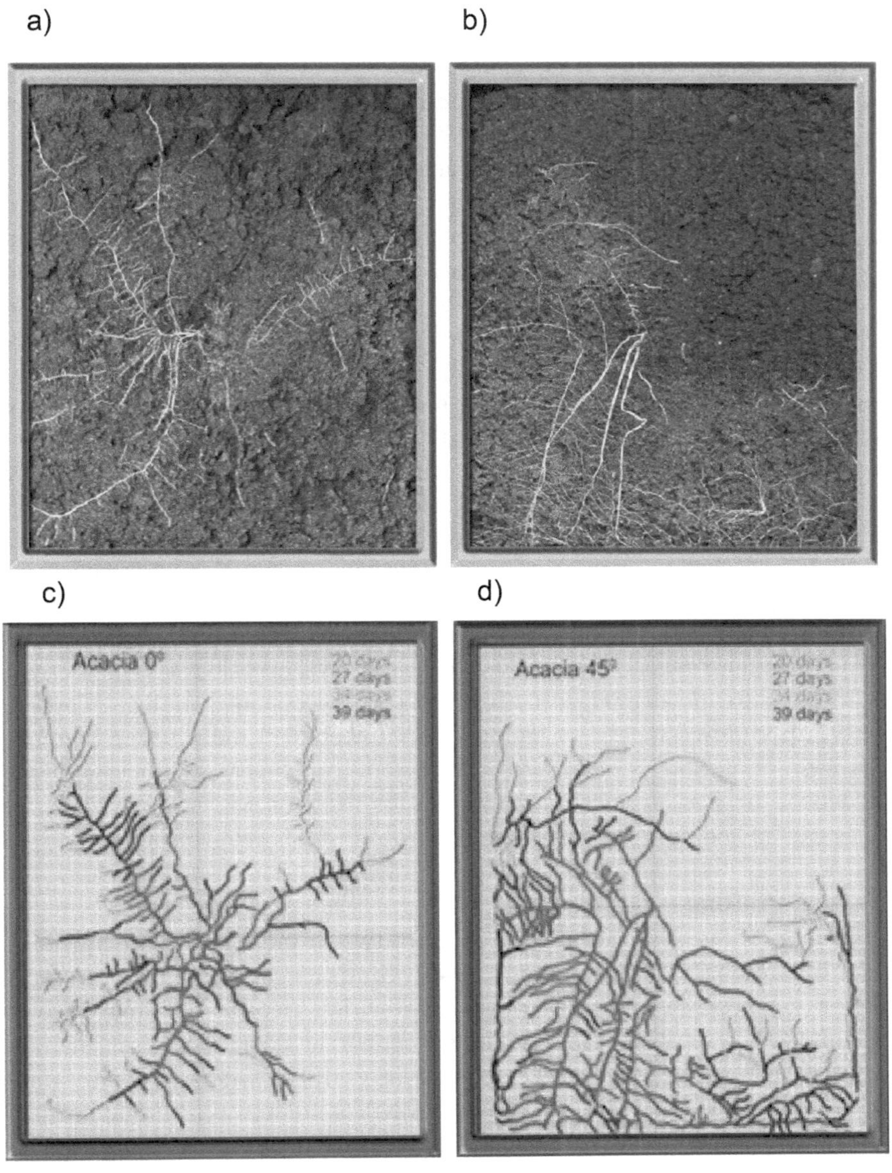

Figure 4-6. Seeds of *Robinia pseudoacacia* germinated in a) a rhizotron inclined at 0° and b) a rhizotron inclined at 45°, showed that initially, roots grew preferentially downhill. Root growth was traced weekly in c) and d) using different colours. In inclined rhizotrons, root growth increased upslope over time, as space and nutrients were exploited downslope (Images courtesy of H. Khuder, (see Khuder 2007)).

that roots clustered uphill, but in a nearby montane forest, roots were predominantly up- or downhill: in both cases, root mass was aligned with prevailing wind direction. Like Nicoll et al. (2006), Soethe et al. (2006) suggested that mechanical loads from prevailing winds had a greater effect on root asymmetry than slope alone. However, in germinating seedlings of *Robinia pseudoacacia*, and *Pinus pinaster* growing in rhizotrons inclined at angles of 22.5° and 45°, Khuder (2007) observed that lateral roots emerging on the uphill side of the taproot changed direction to grow downhill and attributed this reaction to a gravitropic effect (Figure 4.6). This presumed sensitivity to gravity disappeared over time. Once the nutrients in the soil on the downslope side had been exploited, root growth uphill increased. In the field, debris and nutrients will probably accumulate on the uphill side of a tree, thus also encouraging root growth in this zone. If soil is not perfectly stable, it can slide or creep downhill, resulting in tensile stresses in uphill roots. Soil movement is rarely considered in studies of root architecture and should be examined in order to determine its effect on acclimative asymmetric growth of roots. Other mechanisms to be considered include the effects of soil moisture and light on root growth. Coutts and Nicoll (1991) showed that downhill lateral roots could grow out of the soil, but to avoid death by desiccation or in response to light levels, they change direction by bending of the apex back to the deeper soil layers or beneath the soil surface. A similar behaviour has been frequently observed in partly exposed surface roots of mature *Quercus pubescens* growing on steep slopes, which curve right back to the deeper soil layers. As a consequence of this adaptive growth, the downslope root biomass was lower than the upslope biomass (Di Iorio et al. 2005). Therefore, root system asymmetry on slopes appears to depend on age, species and site, and for the moment, no given general rules can be laid down to determine how root systems grow on slopes.

Root grafting

Root grafting is the functional union of two or more roots subsequent to their formation (Kűlla and Lõhmus 1999). Grafts can be found between roots of the same tree, or of roots of a neighbouring tree of the same species (Figure 4.7a). Root grafting is more frequent in deciduous trees than in conifers and not all species are capable of grafting. Grafts generally only form between roots where secondary growth is underway. Formation of a root graft begins due to the mechanical pressure between roots undergoing secondary thickening and are most common in the basal parts of woody roots (Figure 4.7b). When two roots are pressed together during growth, thinning of the bark occurs at the contact surface and proliferating wood cells form a callus until the two roots are joined. Transport of water and nutrients can then pass from one root to another, as well as pathogens. Although some advantages exist in trees where grafting has occurred i.e. the survival of suppressed trees and increased tree stability, it is generally advised to avoid root grafting in plantation forests, thus minimizing the risk of infection of root rot. If root infection is present in a monospecific stand, the most practical method to avoid grafting is to reduce stand density to 2500 stems ha^{-1}, keep a distance of 1.5 – 2 m between trees and complete thinning by the age of 15-20 years (Kűlla and Lõhmus 1999).

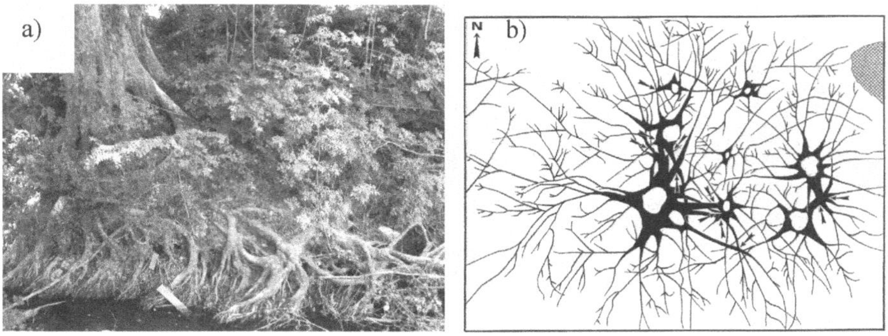

Figure 4-7. a) Photograph of grafted roots in a plane tree (*Platanus acerifolia* Ait.) growing on a river bank. Sediment and debris remain trapped in the root network (Photo: A. Stokes). b) Drawing of root grafts between Norway spruce (*Picea abies* L.) trees in a forest stand. More grafts (arrows) occur nearer the tree stem than at the root apices (Kűlla and Lõhmus 1999, reprinted by permission of the publisher).

Although little work has been carried out on the increase of stand stability due to root grafting, it can be assumed that dense planting will increase the number of grafts. Nonetheless, if trees are linked by a network of root grafts, stability may even be reduced in monospecific stands, due to the "domino" effect of one tree overturning, and bringing its neighbours down at the same time. In a mixed species forest, this problem should not occur.

With regards to slope stability, root grafting should increase soil fixation, by providing a network of roots which can tightly hold the soil in place between roots. Grafts are more common between lateral roots than vertical roots and so are more useful in helping prevent surface erosion than soil fixation in deeper layers. If one tree in the network dies, roots of the living trees will remain attached to the dead and decaying roots and stumps. This interaction is not necessarily detrimental to the health of the living trees (DesRochers and Lieffers 2001).

1.4.3 Root strengths

Root strength varies enormously, not only inter- and intra-species, but also within the same root system, and may depend on the mechanical role of the root. Tensile strength is considered to be one of the most important factors governing soil stabilisation and fixation (e.g. Greenwood et al. 2004), and has therefore been studied in great detail (Hathaway and Penny 1975; Burroughs and Thomas 1977; Schiechtl 1980; Nilaweera and Nutalaya 1999; Genet et al. 2005; Norris 2005b). The tensile strength of roots depends on species and site specific factors such as growing environment, season, altitude and orientation (Gray and Sotir 1996). Root tensile strengths are commonly measured using Universal Testing machines, whereby the root is cut to a required length, clamped into the machine and tested to the point of failure. Tensile strengths for selected European shrub and tree species are summarized in Table 4-3. Tensile strengths vary significantly with diameter, age and method of testing e.g. in a moist or air dry state. The values listed in Table 4-3 should be considered as approximate averages or as a range of values, where values have been found by different researchers they are listed separately. Caution should be applied when using this table, as standard testing procedures do not exist and root diameters are not given.

Root tensile strength is significantly affected by differences in root diameter, as a decrease in strength with increasing root diameter has been well recognised, but this is not a rule for all woody species (Figure 4.8, O'Loughlin and Watson 1979). The variation in root tensile strength with root diameter for several tree species is approximately 8 to 85 MPa for root diameters ranging from 1 to 12 mm (Figure 4.8), but this varies enormously (see Table 4-3). A decrease in root diameter from 5 to 2 mm can result in a doubling or even tripling of tensile strength. This phenomenon has been attributed to differences in root structure, with thinner roots possessing more cellulose than thicker roots, cellulose being more resistant than lignin in tension (Genet et al. 2005). It is not yet known if cellulose content is greater in young roots (which are usually thinner), but initial studies suggest that in conifers, tensile strength is greater in roots from older trees (Genet et al. 2006a).

Other factors which may govern root strength include the mode of planting: naturally regenerated Scots pine had roots more resistant in tension than those of planted Scots pines (Lindström and Rune 1999). The soil environment may also determine root strength: roots of *Zea mays* L. growing in weak soil were stiffer than those growing in strong soil (Goodman and Ennos 1999). The time of year may also be determinant as in temperate regions, roots were found to be stronger in winter than in summer, due to the decrease in water content (Turmanina 1965). In arid regions the opposite may occur. A decrease in tensile strength with increasing altitude has also been found in *Abies georgii* var Smithii. although the mechanism by which this occurs is not yet known (Table 4-3, Genet et al. 2006b).

Contrary to the increase in tensile strength with decreasing root size, compression and bending strength decrease with decreasing root size, this being more pronounced in species with heart- and tap-root systems compared to lateral roots from trees with plate-root systems (Stokes and Mattheck 1996; Stokes and Guitard 1997). Depending on the mechanical role of a root in a system, wood strength will change to resist the forces acting on that root, e.g. leeward roots are more resistant in compression compared to windward roots. This increase in strength probably being due to a greater lignin content (Stokes et al. 1998). In 8 month old Spanish broom (*Spartium junceum*), a significantly higher lignin content was found in root systems growing on slopes compared to those growing on flat ground (Scippa et al. 2006). Root strength may even increase at certain points along a root, in order to resist rupture as that root repeatedly bends during wind sway (Stokes 1999). In trees growing on slopes, tensile strength is greater in upslope roots, compared to downslope and horizontal lateral roots (Schiechtl 1980). Such changes in wood strength may be due to changes in wood anatomy or cellulose content (Khuder 2007), although an extensive study has yet to be carried out.

Table 4-3. Root strengths of shrub and tree species. Most tensile testing was carried out on roots with diameters ranging from 0.5 – 15 mm. Key: σ_T – mean tensile strength (MPa); σ_C – mean compression strength (MPa); σ_B – mean bending strength (MPa); a.s.l. – above sea level.

Author	***Species***	***Common Name***	σ_T	σ_C	σ_B
SHRUB SPECIES					
Mattia et al. (2005)	*Atriplex halimus*	Mediterranean saltbush	57		
Schiechtl (1980)	*Castanopsis chrysophylla*	Golden chinkapin	18		
Schiechtl (1980)	*Ceanothus velutinus*	Ceanothus	21		
Norris (2005a)	*Crataegus monogyna*	Hawthorn	8		
Schiechtl (1980)	*Cytisus scoparius*	Scotch broom	32		
Mattia et al. (2005)	*Pistacia lentiscus*	Gum mastic	55		
Norris and Greenwood (2003)	*Spartium junceum*	Spanish broom	17		
Schiechtl (1980)	*Lespedeza bicolor*	Scrub lespedeza	71		
Norris and Greenwood (2003)	*Phillyrea latifolia*	Privet	11		
Schiechtl (1980)	*Vaccinium* spp.	Huckleberry	16		

TREE SPECIES: CONIFER					
Stokes (unpub. data)	*Abies alba*	Silver fir	31	26	
Riedl (1937)	*Abies brachyphylla*	Nikko fir	28		
Schiechtl (1980)	*Abies concolor*	Colorado white fir	11		
Genet et al. (2006b)	*Abies georgii* 3400 m a.s.l. 4330 m a.s.l.		 28 13		
Genet et al. (2006a)	*Cryptomeria japonica*	Japanese cedar	8-88		
Stokes & Mattheck (1996) Bischetti et al. (2005)	*Larix decidua*	European larch	 66-428	25	5
Schiechtl (1980), Bischetti et al. (2005) Genet et al. (2005) Turmanina (1965); Stokes & Mattheck (1996)	*Picea abies*	European spruce	28 86-650 20-155	 27	 6, 28
Riedl (1937)	*Picea excelsa*	Bhutan pine	28		
Coppin & Richards (1990) Schiechtl (1980) Coutts (1983b) Parr and Cameron (2004) Lewis (1985)	*Picea sitchensis*	Sitka spruce	23 16 35 40	 14-50 	
Schiechtl (1980)	*Pinus densiflora*	Japanese red pine	32		
Norris (unpub. data)	*Pinus halepensis*	Aleppo pine	29, 47		
Schiechtl (1980)	*Pinus lambertiana*	Sugar pine	10		
Genet et al. (2005)	*Pinus nigra*	Austrian pine	10-80		
Ziemer (1981)	*Pinus ponderosa*	Western yellow pine	10		
Genet et al. (2005)	*Pinus pinaster*	Maritime pine	10-132		
Schiechtl (1980)	*Pinus radiata*	Radiata pine	18		

Lindström & Rune (1999) Stokes & Mattheck (1996)	*Pinus sylvestris* - paperpot - natural regeneration	Scots pine	 7 20	 23	 3.5
Schiechtl (1980) Commandeur & Pyles (1991)	*Pseudotsuga menziesii* - Pacific coast - Rocky mountains	Douglas fir	 55 19-61 13, 17		
Schiechtl (1980)	*Tsuga heterophylla*	Western hemlock	20		
Schiechtl (1980)	*Thuja plicata*	Western red cedar	56		
DECIDUOUS					
Schiechtl (1980)	*Acacia confusa*	Acacia	11		
Niklas (1999)	*Acer saccharum*	Sugar maple		35	
Riedl (1937)	*Acer platanoides*	Norway maple	27		
Norris (unpub. data)	*Acer pseudoplatanus*	Sycamore	2		
Schiechtl (1980)	*Alnus firma* var. *multinervis*	Alder	52		
Greenwood et al. (2001)	*Alnus glutinosa*	Common alder	7		
Schiechtl (1980)	*Alnus incana*	Grey alder	32		
Schiechtl (1980)	*Alnus japonica*	Japanese alder	41		
Bischetti et al. (2005)	*Alnus virida*	Green alder	20-92		
Schiechtl (1980)	*Betula pendula*	Silver birch	37		
Stokes & Mattheck (1996) Genet et al. (2005)	*Castanea sativa*	Sweet chestnut	 5-201	24	10
Bischetti et al. (2005)	*Corylus avellana*	Hazel	68-257		
Stokes & Mattheck (1996) Bischetti et al. (2005) Genet et al. (2005)	*Fagus sylvatica*	Common beech	 57-731 40-60	34	15, 32

Riedl (1937) Bischetti et al. (2005) Stokes & Mattheck (1996)	*Fraxinus excelsior*	Ash	26 37-297	26	12
Schiechtl (1980)	*Nothofagus fusca*	Red beech	36		
O'Loughlin & Watson (1979)	*Nothofagus sp.*	Southern beech	31		
Schiechtl (1980)	*Populus deltoides*	Poplar	37		
Schiechtl (1980)	*Populus euramericana*	American poplar	32		
Coppin & Richards (1990) Stokes & Mattheck (1996)	*Populus nigra*	Black poplar	5-12	20	5.5
Hathaway & Penny (1975)	*Populus yunnanensis*	Poplar	41		
Norris & Greenwood (2003)	*Quercus coccifera*	Oak	13		
Riedl (1937)	*Quercus pedunculata*	English oak	45		
Norris & Greenwood (2003)	*Quercus pubescens*	Downy oak	7		
Schiechtl (1980)	*Quercus robur*	English oak	32		
Turmanina (1965)	*Quercus rubra*	Red oak	32		
Norris (2005a)	*Quercus* sp.	Oak	7		
Coppin & Richards (1990) Khuder (2007)	*Robinia pseudoacacia*	Black locust	68 5-32		
Bischetti et al. (2005)	*Salix caprea*	Goat willow	48-409		
Coppin & Richards (1990)	*Salix cinerea*	Grey willow	11		
Schiechtl (1980)	*Salix fragilis*	Crack willow	18		
Schiechtl (1980)	*Salix helvetica*	Willow	14		
Schiechtl (1980)	*Salix matsudana*	Contorted willow	36		
Schiechtl (1980) Bischetti et al. (2005)	*Salix purpurea*	Purple willow	36 51-522		
Schiechtl (1980)	*Sambucus callicarpa*	Pacific red elder	19		
Norris (2005b)	*Sambucus nigra*	Elder	28		
Schiechtl (1980)	*Tilia cordata*	Small leafed lime	26		
Riedl (1937)	*Tilia parvifolia*	Lime	21		

Care must be taken when using this table, as the methodology employed differs between authors. Root diameter is not given and is an important factor when considering root strength (Bischetti et al. 2005; Genet et al. 2005).

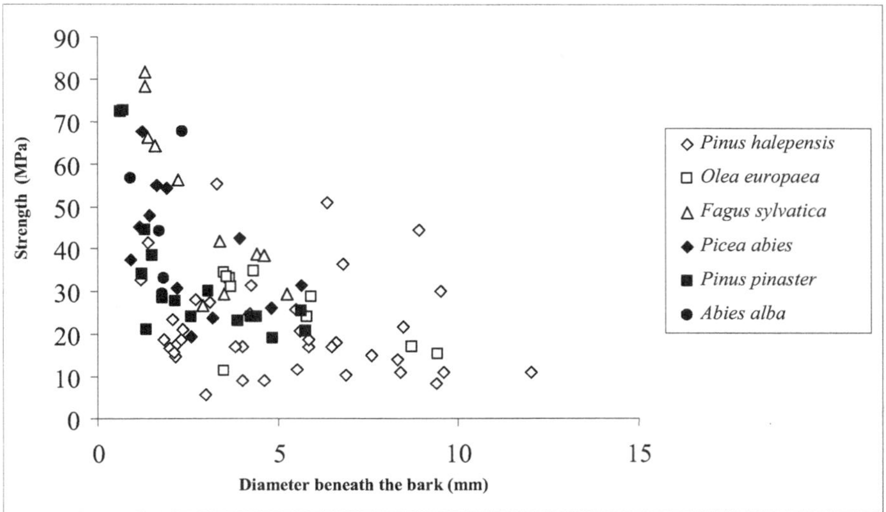

Figure 4-8. Tensile strength decreases with increasing root diameter in several species, except *Pinus halepensis* (from Genet et al. 2005; van Beek et al. 2005; Stokes unpublished data).

Uprooting Strength

Vertical uprooting of whole plants has also been used to determine the contribution of a root system to soil fixation. Although more difficult to quantify and to interpret the results, as roots break during rupture and so the complete architecture is not easy to measure, useful information is nonetheless obtained (Norris 2005a, b). In particular, the force required to uproot herbaceous plants can allow the comparison of several species. For example, vetiver (*Vetiveria zizanioides*) grass, sometimes called the 'living nail' (http://www.vetiver.com) because of its extremely deep and fibrous root system is often used for reinforcing soil on slopes. In 2 year old vetiver plants where mean total dry mass was only 41 g, Mickovski et al. (2005) found a mean uprooting resistance of 467 N. However, in another grass species, big node bamboo (*Phyllostachys nidularia* Munro), mean uprooting resistance was only 1615 N which was very low, considering that mean shoot dry biomass was 359 g (Stokes et al. 2007b). In a similar study of uprooting resistance of several young riparian tree species ranging from 0.6–0.9 m in height and with a shoot dry mass between 20–27 g, Karrenberg et al. (2003) found that uprooting resistance varied between 299–638 N. Similar techniques can be used to determine the uprooting resistance of individual roots and Norris (2005a) found that individual roots of *Quercus robur* L. and *Crateagus monogyna* Jacq. growing on highway slopes required uprooting forces of 3000 to 12000 N to induce failure.

In older trees it is not possible to carry out vertical uprooting tests but overturning tests can provide information about how trees fail when subjected to wind loading, rockfall, avalanches and landslides. The simplest overturning tests use a winch attached via a cable to the trunk of a test tree. The tree is then winched sideways until failure and the force required to uproot or break the tree is measured using a load cell (e.g. Coutts 1983a, 1986; Cucchi et al. 2004; Stokes et al. 2005; Peltola 2006). By calculating the bending moment of the tree (in its simplest form, the force required to cause failure multiplied by the length of the lever arm, which is height to the cable attachment), it is possible to compare several species of different sizes and ages (Peltola 2006), although soil conditions should be taken into account.

It is also possible to obtain useful information about modes of uprooting through the examination of fallen trees in a forest e.g. Abe and Ziemer (1991a, b) reported that most roots broke near their tips where the diameter is less than 1 to 2 cm. This suggests that most roots were pulled out leaving the finest distal portions still embedded in the soil. In general, it has been assumed that roots crossing a shear zone generate tensile strength, are elongated in tension, and break at the tips, not in the shear zone. Thus, the mode of root failure is similar to that occurring during a pull-out test (O'Loughlin and Watson 1979; Abe and Ziemer 1991a, b). The pull-out resistance increases with the number, radius and length of the roots (Abe and Ziemer 1991a, b; Ennos 1993) and can usually be predicted using a combination of eg. volume or number and basal diameter of lateral roots (Bailey et al. 2002; Dupuy et al. 2005b; Stokes et al. 2007b). However, the stiffness of the root material and the soil matric suction might need to be considered in the prediction of root pullout resistance since increased material stiffness contributes to the uprooting resistance of the roots, while increased soil matric potential adds to the effective stress acting on the roots and could increase the uprooting potential of small roots manifold (Mickovski et al. 2007).

In adult trees, the high rate of branching near the stem, or large, rigid main taproot, found in heart and tap root systems, respectively, allows a faster dissipation of forces nearer the stem, therefore a high investment in strength further along the root is not necessary (Ennos 1994; Stokes and Mattheck 1996). However, the stronger the taper, the shorter the lever arm will be (Coutts et al. 1999). Di Iorio et al. (2005) found that in *Quercus pubescens* growing on a hillslope, most of the root biomass was concentrated in several large asymmetrically clustered roots, and that branching points were located further away from the stump. Therefore, the lever arm increased in length thus augmenting the tree's resistance to the turning moment induced by the slope.

On a single root scale, the root uprooting mechanism will vary depending on the stiffness of the root material (Mickovski et al. 2007). Similarly as in other rigid reinforcement materials a rigid woody root will mobilize its pull-out resistance through interface shear equally over the whole length even at very small displacements, offering more resistance to uprooting. In contrast, more flexible roots will tend to mobilise their interface strength progressively with depth while their laterals, once their peak strength is mobilised, will be bent and pulled out together with the vertical root. If the tensile strength of the root is smaller than either soil shear strength or root-soil interface friction, the root will break at the point where the ultimate tensile stress was developed in it without being pulled out (Ennos 1989, 1990).

Mechanisms of uprooting in trees

Most studies of tree uprooting (or overturning) have been carried out with regard to wind storms, and very few have concerned directly mechanical stability on slopes with regard to mass movements (Johnson 1987; Stokes et al. 2005). However, the mechanism of uprooting is similar when overturning forces are either applied more or less along the whole stem length e.g. wind or avalanche forces, or at a single point along the stem e.g. rockfall. With regard to shallow landslides, the root system is sheared due to soil movement and this type of failure is the least similar to failure through wind loading. Therefore, in this next section, we describe uprooting mechanisms with regard to wind loading, but this knowledge can be applied to tree behaviour during avalanches and rockfalls.

When a tree uproots during a wind storm, the mode of failure observed depends largely on the morphology of the root-soil plate and the soil type. During the first stage of uprooting in trees, the weight of the root-soil plate i.e. the roots and adhering soil provide the initial resistance to overturning. If the force on the stem is greater than the resistance of the root-soil plate, the tree will uproot and the soil underneath and around the edge of the plate is broken. The tensile strength of the roots on the windward side of the plate provides high resistance to uprooting, whereas the bending strength of the leeward roots and soil offers a lower resistance (Coutts 1983a, 1986). The contribution of each of these anchorage components will differ depending on the width and depth of the plate (Coutts 1983a). Shallow rooted species uproot at low wind loads, often with the root-plate being completely lifted out of the ground. However, in trees with heart root systems, the root-soil ball slides into the soil. Soil type is also a major factor governing the mode of anchorage. In numerical simulations of overturning of different root system architectures in two different soil types, the root-soil plate was more circular in clay-like soil compared to sandy-like soil (Fourcaud et al. 2007, Figure 4.9). In a similar model using more complex architectures, Dupuy et al. (2005a) showed that rooting depth was a determinant parameter in sandy-like soils, but that overturning resistance was greatest in heart- and tap-root systems whatever the soil type. However, the heart root system was more resistant on clay-like soil whereas the tap root system was more resistant on sandy-like soil. Plate-like systems were the least resistant regardless of soil type.

In trees with deep tap roots, the tree rotates and bends on the windward side of the tap root. The tree can be said to act like a stake, with the taproot the point of that stake (Ennos 1994). The tap root itself pushes into the soil on the leeward side, the top half rotating, and the bottom half remaining reasonably well-anchored. A crevice is then formed on the windward side, becoming larger as the tree is pulled over (Crook and Ennos 1997). Hintikka (1972) also found that the lower half of the tap root may make a semi-circular movement and push into the soil on the windward side. In such a case, the tap root is firmly attached to the soil at its distal end, and the lateral roots hold the stem so rigidly that the tap root has to move in the opposite direction. Trees with well developed taproots, usually do not fail with the tap root slipping out of the ground, as in certain herbaceous species (Ennos 1989). However, the mode of failure does appear to depend on tree age (Cucchi et al. 2004).

Although the mechanism of tree failure during a landslide is different to that during a wind storm, Wu et al. (2004) showed that tree species with the above three types of root systems fail differently in a landslide. A taprooted

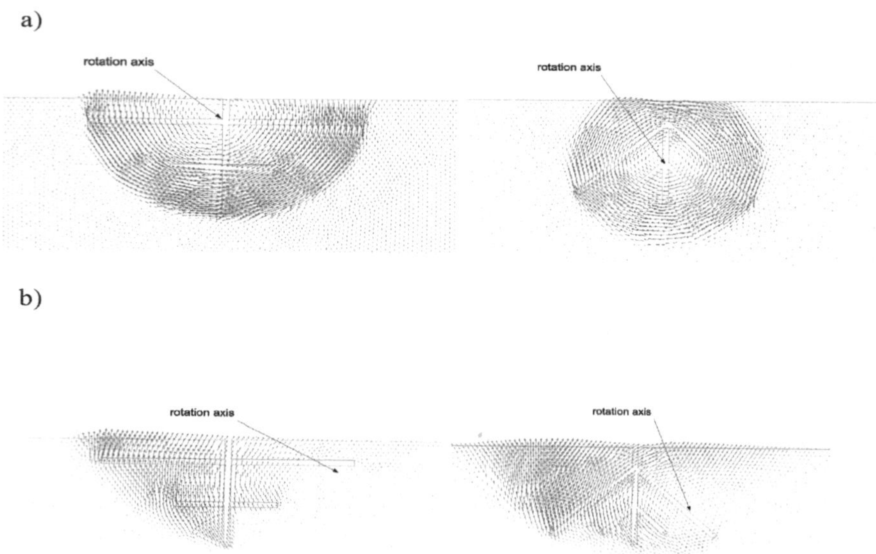

Figure 4-9. Numerical simulations of uprooting of two types of simple root architecture in a) clay-like soil and b) sandy-like soil. The displacement field in soil at the end of the uprooting simulations is shown. The point about which the root system rotates (rotation axis) is shown. The root-soil plate is more circular in clay-like soil compared to sandy-like soil, regardless of root system type (image courtesy of T. Fourcaud, see Fourcaud et al. 2007).

tree will more likely develop the full tensile strength of the taproot, as stresses are concentrated in the one main root. However, in plate or heart rooted trees, many roots do not fail in tension at large shear displacements, as stresses are distributed throughout several roots. Therefore the full tensile strength of all roots in the root system is not utilized. Wu et al. (2004) thus suggested that taprooted trees would be better for stabilizing slopes, as the slope Factor of Safety (see Chapter 5) would be increased.

Seedlings usually possess a tap root and a high root:shoot ratio. In plate and heart systems the tap root dies with age. In most trees, the root:shoot ratio decreases with age. In very old trees, the root system may also have the appearance of a plate system due to its relatively low volume compared to the trunk and crown (Ennos 1994). Therefore, these temporal aspects must also be considered when choosing which species to plant on unstable slopes.

Effect of plant origin on root growth and anchorage

Aside from species and soil conditions, the root development of planted trees is influenced by the planting method, quality of planting and root pruning (undercutting). Three main methods exist when establishing a planted stand: direct seeding on site, transplanting of seedlings sown in containers, planting of bare-root seedlings and transplanting of cuttings (bare-root or in containers). A fundamental difference between seedlings and cuttings is that the latter do not have a tap-root, but can develop one after about five years (Figure 4.10; Khuder et al. 2007). It is generally considered that naturally regenerated and direct sown seedlings are the most mechanically stable and more difficult to uproot (Halter and Chanway 1993; Lindström and Rune 1999). This stability is probably due to a well-developed and undisturbed root system. Container grown seedlings often have a limited root system, with lateral roots spiralling around the container (Lindström and Rune 1999), although several types of container now exist with slits whereby lateral roots can grow through the slit (Rune 2003). Bare-root seedlings are often deformed during transplanting and roots damaged or

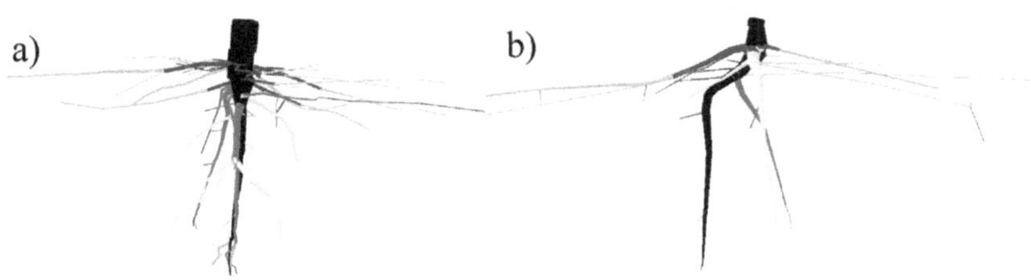

Figure 4-10. Differences in root architecture occur depending on the plant material used. Reconstruction of root systems coloured as a function of compartment type in a) a Maritime pine sapling planted as a paper pot seedling and b) cutting of the same species where a lateral root has grown downwards and acts as a taproot after 7 years growth (Khuder et al. 2007; reprinted by permission of the publisher).

bent (Nörr 2003). Trees generated from cuttings are usually smaller with a lower number of roots than trees grown from seeds.

Cuttings do not have the same ability to generate lateral and vertical roots, at least in young trees (Figure 4.10). Cuttings are easier to uproot than seedlings the same age, but these differences may disappear after several years (Khuder et al. 2007). Roots from naturally regenerated trees are thought to have a higher tensile strength than container plants (Lindström and Rune 1999), whereas no differences have yet been found between cuttings and container grown seedlings.

1.4.4 Cohesion and root reinforcement

Roots of vegetation are known to stabilize, or, improve the bearing capacity of soils on which they grow. Evidence of this has been reported in forest soils (Wasterlund 1989; Makarova et al. 1998), slopes (Waldron 1977; Waldron and Dakessian 1981, 1982; Terwilliger and Waldron 1991). Investigations conducted by Willatt and Sulistyaningsih (1990) on loamy soil showed increases in both bearing capacity and shear vane resistance in the presence of roots, whilst Goss (1987) reported an increase in the soil bulk density in similar studies. According to Wasterlund (1989), the increase in soil strength caused by the presence of tree roots may range between 50 and 70%.

The intermingled roots of plants tend to bind the soil together in a monolithic mass and contribute to strength by providing an additional apparent cohesion, c_R (see Gray and Leiser 1982; Chapter 5). As a result of their random orientation, roots have a negligible influence on the frictional component of soil strength. Thus, in a root-permeated soil the Mohr-Coulomb failure criterion is modified to include c'_R (effective root cohesion):

$$s = c' + c'_R + (\sigma - u) \tan\phi' \tag{4}$$

where s is the shear strength of the soil-root composite, c' is effective cohesion, σ is normal stress, u is pore-water pressure and ϕ' is the effective angle of internal friction. The magnitude of c_R varies with the distribution of the roots within the soil and with the tensile strength of individual roots (Wu et al. 1979).

1.4.5 Surcharge

Tree surcharge is the weight of an individual tree on the slope, or when viewed in a slope context, the combined weight of all vegetation. This weight depends on species, diameter, and height. A whole forest on a slope represents a relatively small surcharge when compared to soil mantle and other weight factors (Greenway 1987). The surcharge, or overall load contributed to the slope by vegetation, is not seen as having a significant influence on slope stability (Greenway 1987; Greenwood et al. 2004).

The additions of extraneous forces on vegetation contribute to the overall weight of the vegetation structure on the slope. The weight of snowfall on the canopy of a tree, for example, increases its weight force, as well as providing additional loads on branches. These forces can be considered as static forces, the effects of which may be transferred (to a very limited extent) into the slope.

Vegetation surcharge increases normal and downhill weight force components on potential slip surfaces. If the slope angle is greater than the angle of internal friction, a stabilising influence results (Gray and Megahan 1981). The model developed by Gray and Megahan (1981) demonstrates that surcharge is beneficial when the following equation is satisfied (Equation 5).

Surcharge relationship:

$$c < \gamma_w H_w \tan \phi \cos^2\beta \tag{5}$$

where c = cohesion
γ_w = unit weight of water
H_w = groundwater height above slip plane
ϕ = angle of internal friction
β = slope/slip plane angle.

The equation demonstrates that surcharge may be beneficial to infinite slopes when cohesion is low, groundwater level and soil friction values high and the slope angle is relatively low (Greenway 1987). Nevertheless, surcharge usually has a small effect on slope stability analyses and even after clear-felling of a forested slope, increases in vegetation surcharge are assumed to be slightly lower than the recovery of rooting strength (Sidle 1992; Dhakal and Sidle 2003).

1.4.6 Buttressing and arching

Trees with stems and root systems of sufficient girth block soil movement simply due to their presence, in a phenomenon known as buttressing. During buttressing, a cylinder of soil upslope of the tree is stabilised, and exerts a static force on the stem. This force may increase incrementally over time, as more surface slope material is gradually buttressed. Given certain spacing between neighbouring tree stems, arching may also subsequently develop. Arching is a condition where soil is stabilised between two buttresses.

The combined forces exerted on tree stems and surface root systems as a result of both buttressing and arching are considered as static forces, due to the slow nature with which they incrementally increase in magnitude. They are derived from the downhill component of the weight force of the soil. The quantity of soil build-up in the area behind the tree is dependent on erosive processes as well as soil movement in a shear zone. In a situation where arching develops, trees growing on a slope can be said to act like piles, anchored into a firm subsurface strata (Gray and Megahan 1981).

Wang and Yen (1974) developed a model for arching on slopes using theory based on a semi-infinite slope model using a condition of rigid plastic solid soil behaviour. The model assumes a single row of trees ('embedded piles') of a given diameter and spacing on a slope (Figure 4.11). The total force (P) against a pile embedded in a slope is given in Equation 6.

Pile force:

$$P = K_o/2\ \gamma H^2 d + (K_o/2\ \gamma\ H - p)BH \qquad (6)$$

where P = force on pile (tree stem)
K_o = coefficient of lateral earth pressure at rest
γ = unit weight of soil
H = soil mantle depth
d = pile diameter
B = clear spacing or opening between piles
p = average lateral pressure or arching pressure.

The load on a pile in this situation effectively involves two loads. Firstly, the load due to the soil pressure uphill of the pile, and secondly, soil arching pressure transferred to adjacent piles similar to a pressure exerted as if each pile is the abutment of an arch dam (Gray and Megahan 1981). The model demonstrates that as P tends to zero, arching action is maximised.

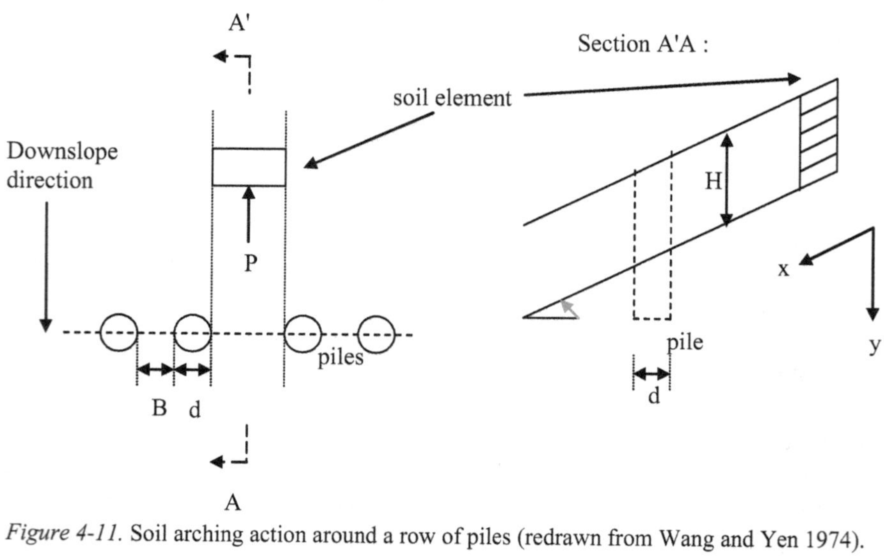

Figure 4-11. Soil arching action around a row of piles (redrawn from Wang and Yen 1974).

Gradual surface soil creep over long timescales may exert an incremental pressure on tree stems which contributes to the phenomenon described above. The resulting tree stabilising mechanisms are stem thickening and buttress formation.

1.5 Models

1.5.1 Hydrological models

Hydrological processes relevant to erosion and mass movement activity can be modelled in various ways. A short overview is given here of the most common modelling concepts. These concepts relate to mathematical models that nowadays supersede analogue and physical models almost completely. Some references are included where one could find more detailed information on specific models.

Hydrological models can be classified according to the characteristics of Table 4-4. Generally, model sophistication increases from left to right. With increasing computational power numerical and physically based models are becoming widely available. These models are mainly dynamic and distributed and sometimes embedded in geographical information systems (GIS), which facilitates the incorporation of spatial information. Such models are both freely and commercially available. These models are widely applicable but uncertainty and data availability are becoming more and more the limiting factor and simpler models are equally useful in many cases e.g. tank models in the case of landslides that describe the water balance of landslides including that of a single leaky reservoir.

Table 4-4. Classification of model approaches

Issue	***Approach***	
Time	Static	Dynamic
Space	Combined	Distributed
Process understanding	Black box	Conceptual
Process representation	Empirical	Physical
Degree of reduction	Deterministic	Stochastic

Erosion and slope stability problems are related to the same hydrological processes but are often approached with different modelling concepts. Erosion is directly related to infiltration and runoff at the soil surface whereas landslides are triggered by the rise in pore pressure deeper in the soil. This makes these problems in essence two-dimensional and erosion problems are represented in plan whereas landslides are represented in profile. If necessary, the neglected direction can be introduced at a reduced level, for example to simulate convergence or divergence in groundwater flow at landslides or to mimic the effect of saturation excess infiltration in erosion problems (2½-D problems).

Hydrological models of erosion are usually built around an infiltration module that controls rainfall excess and a routine equation that describes surface runoff (e.g. Manning's equation). Those of landslides usually describe water flow through the soil by means of Darcy's Law or, as an extension, Richard's equation when flow in the unsaturated zone is considered. Research has shown that water transport through the unsaturated zone is mainly 1D in slopes, unless clear heterogeneity or anisotropy exists in the slope which favours 2D water flow. Such heterogeneities are macropores that with preferential flow may be extremely important in the temporal response of pore pressures and pipe erosion, cannot be fully described physically. A conceptual approach is often followed, representing macropores as separate conduits that exchange water with the surrounding matrix (e.g. Van Beek and Van Asch 1998; Van Asch et al. 2001). In those cases, macropore flow cannot only account for short-circuiting the percolation with the groundwater but also for the increased rate of lateral discharge. Alternatively, preferential flow can be described by a dual permeability function for the matrix (Van Genuchten), which confines the problem to the domain of the Richards' equation. Even simpler approaches account for macropore flow by adding a fraction of the net rainfall directly to the groundwater store (Van Beek 2002; Malet et al. 2003).

1.5.2 Mechanical root reinforcement models

Perpendicular and inclined root reinforcement models

Wu (1976) developed a root reinforcement model for perpendicular roots on a shear plane. Roots, in nature, may act at any angle to the shear plane; therefore, the inclined root reinforcement model (Figure 4.12) was introduced by Gray and Leiser (1982). Both models are limited by assumptions regarding tensile strength and anchorage. The models assume that roots increase soil shear strength and that the magnitude of increase depends on the total area of roots present and the tensile strength of those roots.

The simplified perpendicular root-soil model allows quantification of increased shear strength of soil due to root reinforcement. The mobilisation of the tensile resistance of roots can be modelled as an increase in the shear strength of the soil (ΔS), i.e.

$$\Delta S = t_r (\cos\theta \tan\phi + \sin\theta) \tag{7}$$

where ΔS = shear strength increase from root reinforcement, kPa
θ = angle of intersection with shear zone
ϕ = angle of internal friction
t_r = average tensile strength of root per unit area of soil, kPa.

The average tensile strength of roots per unit area of soil is:

$$t_r = T_r (A_r/A) \tag{8}$$

where T_r = average tensile strength of root (kPa) and A_r/A = root area ratio (RAR) or fraction of soil cross-sectional area occupied by roots.

The angle of root intersection with the shear plane θ, varies with the thickness of the shear zone (Z) and the amount of shear displacement (x) (Figure 4.12). Waldron (1977) and Wu et al. (1979) report that θ varies between 45 and 70°. Tests have shown that it is sufficient to use the simplified perpendicular model for root reinforcement estimates of inclined roots, but Danjon et al. (2007) showed that it is also possible to use true angles of woody roots crossing the potential slip surface.

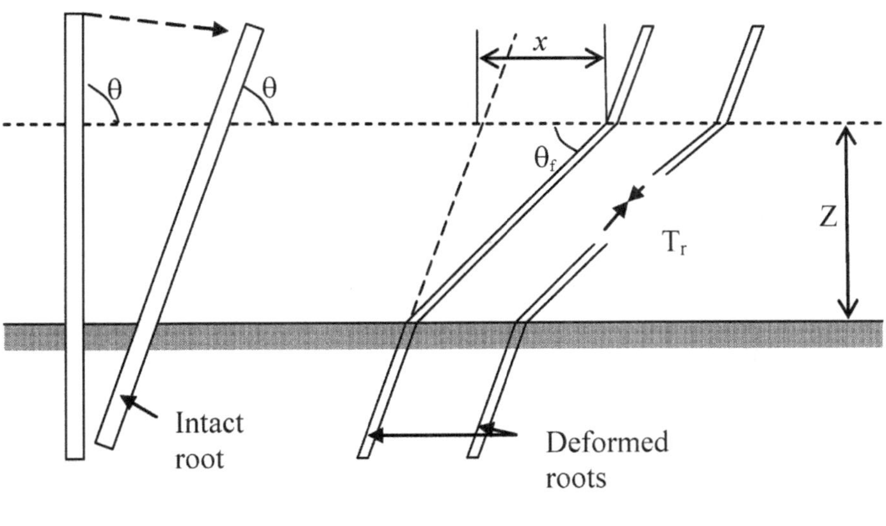

Figure 4-12. Root reinforcement model for perpendicular and inclined roots (modified from Gray and Leiser 1982). Z = shear zone, x = shear displacement, θ = initial angle of intersection with shear plane, T_r = tensile strength of root, θ_f = angle of intersection after deformation.

The models are based on the full mobilisation of the tensile strength of the roots, therefore pull out or bond failure must be prevented. The roots must have sufficient root length beyond the failure zone and sufficient roughness so that the root-soil bond exceeds the tensile strength of the root. Pull out or breaking in tension before the full tensile strength is exacerbated in saturated soils and with fine roots. The minimum length (L_{min}, mm) of roots of uniform thickness (d, mm) required to prevent pull out or bond failure is therefore:

$$L_{min} > \frac{T_R d_R}{2\tau_R} \tag{9}$$

where T_R = tensile strength of root (kPa) and τ_R = maximum bond stress or pull out resistance between root and soil (kPa).

Root stretching occurs when there is insufficient root elongation and constraint to mobilise the root tensile or breaking strength. The mobilised tensile strength of stretched roots (t_{RS}) is determined by the amount of root elongation and the root tensile modulus E_R (Gray and Barker 2004). The mobilised tensile stress (t_{RS}) per unit area of soil is (Waldron and Dakessian 1981):

$$(t_{RS}) = (4z\, \tau_b\, E_R/d)^{1/2}\, (\sec\theta - 1)^{1/2}\, (A_R/A) \tag{10}$$

where z = thickness of the shear zone
τ_b = root-soil bond stress
E_R = tensile modulus of the root
d = root diameter
θ = angle of shear distortion
A_R/A = root area ratio.

The root-soil bond stress can be estimated from the confining stress acting on the roots and the coefficient of friction. For vertical roots, bond stress varies with depth, and is given by the equation:

$$\tau_b = z\,\gamma\,(1 - \sin\phi)\, f \tan\phi \tag{11}$$

where z = depth below the ground surface
γ = soil density
ϕ = angle of internal friction
f = coefficient of friction between the root and soil (varies between 0.7-0.9 for wood and soil) (Gray and Barker 2004).

The increase in shear strength from mobilisation of root tensile resistance from stretching is:

$$\Delta s = (4z\, \tau_b\, E_R/d)^{1/2}\,(\sec\theta - 1)^{1/2}\,(A_R/A)\,(\sin\theta + \cos\theta\, \tan\phi) \tag{12}$$

Equation 12 can be rewritten as:

$$\Delta s = k\, \beta\, (A_R/A)\,(\sin\theta + \cos\theta\, \tan\phi) \tag{13}$$

where $k = (4z\, \tau_b\, E_R/d)^{1/2}$ and $\beta = (\sec\theta - 1)^{1/2}$.

Hence, for the average tensile strength of roots per unit area of soil, typical values of root tensile strengths can be found in Table 4-3 and root densities (RAR) may vary from 0.14 – 5.0% for *Quercus alba* L. (Danjon et al. 2007), 0.10 – 0.35% for *Larix decidua* Mill., *Fagus sylvatica* L. and *Picea abies* L. on silt with clayey sand (Bischetti et al. 2005). In mixed natural forests of the Oregon coast range, the mean RAR ranged between 0.1 and 1% in 1.2 m deep pits dug midway between neighbouring trees (Schmidt et al. 2001), whereas Abernethy and Rutherfurd (2001) found values of 0.001 – 0.756% in *Eucalyptus camaldulensis* Labill. and *Melaleuca ericifolia* Smith. growing along riverbanks in Australia. Therefore, values of RAR are highly variable and particularly susceptible to the effects of larger roots.

Fibre bundle model

The perpendicular root reinforcement model as described above assumes that all of the tensile strength of the roots is mobilized instantaneously at the moment of slope failure. When slopes fail, the root-soil matrix shears, and the roots contained within the soil have different tensile strengths and thus break progressively, with an associated redistribution of stress as each root breaks (Pollen and Simon 2005). This mode of progressive failure is well documented by fiber bundle models in material science (e.g. Callister 2007).

Pollen and Simon (2005) and Pollen (2006) applied the fibre bundle model to root reinforcement of riparian vegetation on streambanks. The fiber bundle reinforcement method uses the concept of global load sharing where a bundle of roots with known number, size and material properties resist the shear force applied to the root-soil composite. To calculate the response of each sample, an initial shear force is applied to the bundle and assumes that each root in the bundle is able to resist an equal portion of the applied force. Since roots in the bundle differ in diameter, the shearing force induces different stress in each root. If the stress induced is higher than the maximum tensile strength of the root, the root is considered as broken and the force it is not able to resist is redistributed to the remaining number of roots in the bundle. This procedure continues iteratively until all of the roots in the bundle are broken or the redistributed force is higher than the force any of the roots were able to withstand. This approach yields root reinforcement lower than the one calculated by Wu's (1976) model.

Energy approach model

The energy approach model was developed by Ekanayake et al. (1997), and Ekanayake and Phillips (1999a,b, 2002), and takes into account the fact that roots can withstand large-strains during displacement of the soil-root system. The characteristics of the shear stress–shear displacement curve obtained from an *in situ* direct shear test are used to find the total energy capacity of the soil-root system and the amount of energy exchanged (see Chapter 5, Figure 5.9 for more details). The energy exchanged during the shearing process is directly related to the area between the stress-displacement curve and the *x*-axis. The total energy capacity of the soil-root system is the area under the soil with roots up to the shear displacement at peak shear stress.

Numerical methods to calculate root-soil mechanical interaction

Using numerical methods to investigate root-soil mechanical interaction can be very helpful if it is necessary to quantify the effect of vegetation on slope stability. These methods are based on a discrete representation of the system mechanical equilibrium that can be solved using a computer. Two approaches can be considered; (1) one consisting of the direct calculation of forces and moments from Newton's second law, i.e. the net force and net moment on every body in an equilibrated system is zero; (2) the second considering the calculation of displacements or velocities of a finite number of points (nodes) of the studied body from the equilibrium equations or equations of motion.

The Limit Equilibrium Method (LEM)

The Limit Equilibrium Method (LEM) is commonly used in geotechnical engineering to estimate the slope factor of safety (FOS, see Chapter 5) in 2D analyses of slope stability, even if it can be applied to 3D situations. The principle is to split the cross section of a slope into slices and to write the equilibrium of forces and moments at the interfaces. The slices are limited by arbitrary vertical cutting planes, the soil surface and the slip surface that is defined a priori, i.e. as input data. The FOS is therefore calculated as the ratio between the shear strength, usually provided by the Mohr-Coulomb failure criterion, and the actual shear force that applies at the slip surface. This method is easy to implement and the calculation is very fast. The impact of vegetation on slope stability can be investigated considering the additional cohesion provided by roots as given by Wu (1976) for example (see section 1.5.2). An adaptation of the LEM taking into account root reinforcement has been proposed by Greenwood (2006; see Chapter 5).

The Finite Difference Method (FDM) and Finite Element Method (FEM)

Alternatively to LEM, more sophisticated and accurate methods can be used to carry out numerical analyses of root-soil interactions. This is the case of the Finite Difference Method (FDM) and the Finite Element Method (FEM), which are both based on a spatial discretization of the studied domain (root-soil medium) that aims in reducing the continuum field functions, e.g. force, displacement, stress or strain, to their values at particular points (nodes). In such displacement approaches, the first stage of the procedure consists of the calculation of nodes' displacement or velocity with regard to the forces applied on the body. In FEM (Zienkiewicz and Taylor 1998), the displacement or velocity field is derived from the integral formulation of the Virtual Work Principle (VWP) or Virtual Power Principle (VPP) respectively. The VWP expresses the equality between the work of external forces and the work of internal forces, or strain energy, for a virtual displacement field. The VPP is based on a similar formulation but introducing virtual velocities and virtual strain rates. In both FDM and FEM, strains or strain rates are expressed as the derivative of the displacement or velocity components, thus providing the strain-displacement or strain rate-velocity relationships. Once the strains or strain rates are deduced from the computed displacements or velocities, constitutive laws, i.e. stress-strain relationships, allow the stress field to be calculated.

Contrary to the LEM where the slip surface is given at the beginning of the slope stability analysis, FDM or FEM allows evolution in time of the system to be simulated and the slip surface location to be derived from the shear stress calculation depending on the considered plasticity criterion. Computation of the FOS can be done using the Shear Strength Reduction technique (SSR) (Zienkiewicz et al. 1975).

Due to their ability to solve very complex problems with a high degree of accuracy, i.e. considering complex geometries and non-linear constitutive laws, these numerical methods are commonly employed in engineering mechanics and physics. They are also becoming more and more popular in geotechnical engineering even though LEM is still the most used method in this field (Duncan 1996; Cai and Ugai 1999; Griffiths and Lane 1999). Few recent research studies have been carried out in the field of ecotechnology, using such approaches to study how vegetation reinforces soil on slopes. Frydman and Operstein (2001) applied the FDM on shear tests of rooted soils at the plant scale using the FLAC software (Itasca 1993). This study demonstrated the ability of the method to solve the problem with an acceptable precision. Other examples of using FDM at the slope scale with consideration of additional cohesion provided by plant roots can be found in Operstein and Frydman (2002), or van Beek et al. (2005). 3D FEM analyses have been recently performed on heterogeneous afforested slopes by Kokutse et al. (2006) who aimed at studying the effect of forest structures and root shape distribution on slope stability (Figure 4.13). Such FEM root-soil analyses were also developed at the plant level using structural beam elements included in a 3D soil medium, in order to study tree uprooting mechanisms considering different root architectures in different soil types (Dupuy et al. 2005a; Dupuy et al. 2007). Other simpler plane strain models (2D models) allowed an understanding of the main components of tree anchorage at the local level (Dupuy et al. 2005b; Fourcaud et al. 2007).

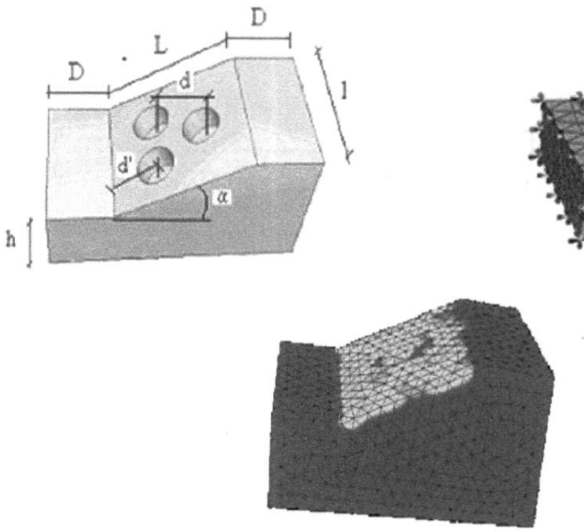

Figure 4-13. Example of 3D FEM model of slope stability with tree root inclusions (image courtesy of NM Kokutse, see Kokutse et al. 2006).

2. WHAT IS THE BEST TYPE OF ROOT SYSTEM FOR STABILIZING SOIL ON A SLOPE?

Once the type of instability process on a slope has been determined e.g. water or wind erosion, shallow landsliding, avalanche or rockfall etc, the type of plantation or management can then be envisaged. On slopes subjected to frequent wind storms, it is also necessary to take into account wind direction and intensity, and use suitable material for withstanding this extra abiotic stress.

Styczen and Morgan (1995) first attempted to classify root systems according to their suitability for stabilizing soil on slopes or their erosion-reducing potential. Types H- and VH included root systems with horizontal lateral roots and deep taproots, respectively (Figure 4.14). M-type root systems have profusely branching roots in the topsoil, but with a narrow lateral extent (Figure 4.14). This type of classification is simplistic but provides a good base on which to develop future research about the use of different species for soil fixation depending on their root architecture (Reubens et al. 2007). In recent years, several studies have provided indicators about root architecture and how it influences slope stability. Where wind erosion is the major cause of soil loss, it is better to choose species with shallow but very dense root systems. Rhizomatous species can be envisaged e.g. bamboo (Storey 2002; Stokes et al. 2007b) as well as clumping grasses and bushy shrubs. The latter types of plants will also help 'capture' soil in their aerial parts. However, the soil fixing characteristics of root systems decrease rapidly with distance from the main plant axis, therefore, where vegetation is patchy, local soil slippage or erosion may occur between plants, especially in clumping species (Terwilliger and Waldron 1991; Danjon et al. 2007; De Baets et al. 2007).

With regard to water erosion, it is important to determine the type of erosion encountered. For splash and interrill erosion, aboveground vegetation cover is the most important vegetation parameter and erosion can be reduced by planting e.g. *Rosmarinus* species which provides good ground cover (Bochet et al. 2006). However, for rill and ephemeral gully erosion, plant roots are at least as important as aboveground cover (Gyssels et al. 2005). In general, dense, lateral spreading root systems would be most useful in fixing soil against rill and gully erosion. De Baets et al. (2007) found that grasses

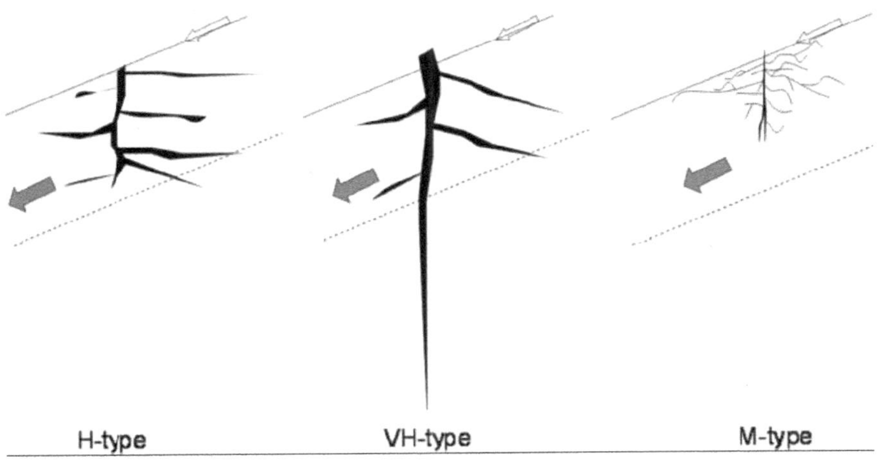

Figure 4-14. Representation of different root classes as identified by Styczen and Morgan (1995). Soil movement (grey arrow) and runoff (white arrow) are indicated. The potential slip surface is indicated by the dashed line. H-type root systems have >80% biomass in the top 0.6 m of soil with roots having a wide lateral extent. VH-type systems have long, thick tap roots and M-type systems have 80% of the root matrix in the top 0.3 m and a narrow lateral extent (Reubens et al. 2007, reprinted by permission of the publisher).

had the highest erosion-reducing potential in situations where overland flow was severe. The grasses examined had a high density of fine roots in the top 0–0.2 m soil. However, this erosion-reducing effect decreased very rapidly with increasing soil depth. Species such as the Mediterranean grass *Stipa tenacissima* L. which have both good ground cover and high root density are therefore highly useful against water erosion in general (De Baets et al. 2007, see Chapter 6).

When reinforcing soils against shallow slope instability, some of the most important criteria to consider are the number, diameter and tensile resistance of roots crossing the slip surface (Greenwood et al. 2004; Cammeraat et al. 2005; van Beek et al. 2005). Therefore, root systems composed of deep taproots and sinker roots crossing the slip surface would be ideal. As root tensile strength is greater in thin woody roots, a large number of small diameter roots would provide a root-soil matrix that resisted shear better. Vetiver grass is often used for replanting on shallow slope failures, due to its deep and fibrous root system, which can cross the slip surface (if the slip surface does not include bedrock). However, at the top or toe of a slope, it would also be necessary to have roots crossing the vertical slip surface in order to prevent slope failure. Horizontal lateral roots are therefore also necessary as they can provide lateral reinforcement (Zhou et al. 1997). Thus, the ideal root morphology in shrubs and trees would be a heart root system, with deep sinkers and wide-spreading lateral roots. Nevertheless, at the centre of a slope, taprooted species could be planted, as the slip surface would most likely be parallel to the soil surface (assuming that slope failure did not occur in the middle of the slope). However, in the middle of many cut-slopes and embankments, the slip surface is most likely to be circular at a depth of 1.5-2.0 m (Perry 1989), therefore the root network (tap or heart system) should have sufficient depth to interact with the slip surface. Initial stabilization may be achieved by using 2.0 m long willow poles inserted on a regular spacing across the slope, which will then over time sprout roots at the required depth to maintain stability over the long-term (e.g. Steele et al. 2004). Perry et al. (2003a, b) give advice about vegetation management on infrastructure slopes which should be followed for use in the UK.

In active rockfall corridors, mechanical properties of stem wood are more useful than root system morphology for determining tree resistance to rockfall (Stokes et al. 2005, 2007a; see Chapter 7). Nevertheless, if trees are well-anchored with a deep taproot e.g. *Abies alba*, they will be less likely to uproot when hit by a falling rock, compared to trees with superficial root systems e.g. *Picea abies*. Similarly, tree resistance to windthrow will be enhanced if trees have deeper root systems (see Chapters 6,7). As far as the authors know, no research has been carried out on the performance of different trees species possessing different types of root systems with regard to avalanche resistance. However, the methodology used for determining tree resistance to windthrow and rockfall can also be applied when investigating resistance to avalanches e.g. Johnson (1987) found that on subalpine mountain slopes where avalanches occur, a plant will bend when impacted by an avalanche. If flexible enough, it will deflect and suffer less damage, but if too rigid and unable to bend, will rupture in the stem or uproot. Therefore, well-anchored plants with a low bending stiffness will better survive the passage of an avalanche (Johnson 1987; Kajimoto et al. 2004).

In conclusion, a mixture of species of different ages will usually improve soil fixation. Native species are often a suitable choice as they are already adapted to the local environment. Grasses stabilize the topsoil against erosion and shrubs and trees fix deeper soil, especially if roots can cross the slip surface. If only one species is used e.g. even aged monospecific stands of trees are planted on unstable slopes, it is likely that soil reinforcement will be poor during the early years. Once the trees are established, slope stability will be increased, but if managed incorrectly, e.g. if thinned extensively thus leaving large gaps between trees, unstable zones may form between trees. Similarly, the spatial distribution of vegetation may lead to localized zones of slippage or erosion, and further research needs to be carried out to determine the best pattern for planting trees and shrubs on slopes, depending on the instability process underway (Schmidt et al. 2001; Sakals and Sidle 2004; Kokutse et al. 2006; Danjon et al. 2007).

3. REFERENCES

Abe K, Ziemer RR (1991a) Effect of tree roots on shallow-seated landslides. USDA Forest Service, Gen. Tech. Rep. PSW-GTR-130

Abe K, Ziemer RR (1991b) Effect of tree roots on a shear zone: modelling reinforced shear strength. Can J Forest Res 21:1012-1019

Abernethy B, Rutherfurd ID (2001) The distribution and strength of riparian tree roots in relation to riverbank reinforcement. Hydrol Proc 15:63-79

Allen RG, Pereira LS, Raes D, Smith M (1998) Crop evapotranspiration; guidelines for crop water requirements. FAO Publication 56

Bailey PHJ, Currey JD, Fitter AH (2002) The role of root system architecture and root hairs in promoting anchorage against uprooting forces in *Allium cepa* and root mutants of *Arabidopsis thaliana*. J Exp Bot 53:333-340

Barij N, Stokes A, Bogaard T, van Beek LPH (2007) Does growing on a slope affect tree xylem structure and water relations? Tree Physiol 27:757-764

Bastiaanssen WGM (2000) SEBAL based sensible latent heat fluxes in the irrigated Gediz Basin, Turkey. J Hydrol 229:87-100

Bastiaanssen WG, Meneti M, Feddes RA, Holtslag AAM (1998) A remote sensing surface energy balance algorithm for land (SEBAL), 1. formulation. J Hydrol 212-213:198-212

Betson RP (1964) What is watershed runoff? J Geophys Res 69:1541-1552

Beven K, Germann P (1982) Macropores and water flow in soils II. Water Res Res 18:1311-1325

Bischetti GB, Chiaradia EA, Simonato T, Speziali B, Vitali B, Vullo P, Zocco A (2005) Root strength and root area of forest species in Lombardy. Plant Soil 278:11-22

Bochet E, Poesen J, Rubio JL (2006) Runoff and soil loss under individual plants of a semi-arid Mediterranean shrubland: influence of plant morphology and rainfall intensity. Earth Surf Proc Land 31:536-549

Brutsaert W (1982) Evaporation into the atmosphere. Dordrecht, Kluwer

Burroughs ER, Thomas BR (1977) Declining root strength in Douglas-fir after felling as a factor in slope stability. USDA Forest Service Research Paper INT-190, 1-27

Cai F, Ugai K (1999) 3D FE-analysis of the stability of slope reinforced with piles. Numerical models in geomechanics – NUMOG 7:541-546

Callister WD Jr (2007) Materials Science and Engineering: an Introduction. 7th ed. John Wiley & Sons, Inc., New York

Cammeraat LH (2002) A review of two strongly contrasting geomorphological systems within the context of scale. Earth Surf Proc Land 27:1201-1222

Cammeraat LH, Van Beek R, Kooijman A (2005) Vegetation succession and its consequences for slope stability in SE Spain. Plant Soil 278:135-147

Chiatante D, Sarnataro M, Fusco S, Di Iorio A, Scippa GS (2003) Modification of root morphological parameters and root architecture in seedlings of *Fraxinus ornus* L. and *Spartium junceum* L. growing on slopes. Plant Biosyst 137:47-56

Commandeur PR, Pyles MR (1991) Modulus of elasticity and tensile strength of Douglas fir roots. Can J Forest Res 21:48-52

Coppin NJ, Richards IJ (1990) Use of Vegetation in Civil Engineering. CIRIA, Butterworths, London

Coutts MP (1983a) Root architecture and tree stability. Plant Soil 71:171-88

Coutts MP (1983b) Development of the structural root system of Sitka Spruce. Forestry 56:1-16

Coutts MP (1986) Components of tree stability in Sitka spruce on peaty gley soil. Forestry 59:173-197

Coutts MP, Nicoll BC (1991) Orientation of the lateral roots of trees. I. Upward growth of surface roots and deflection near the soil surface. New Phytol 119:227-234

Coutts MP, Nielsen CCN, Nicoll BC (1999) The development of symmetry, rigidity and anchorage in the structural root system of conifers. Plant Soil 217:1-15

Crook MJ, Ennos AR (1997) The increase in anchorage with tree size of the tropical tap rooted tree *Mallotus wrayi*, King (Euphorbiaceae). In: Jeronimidis G, Vincent JFV (eds) Plant Biomechanics. Centre for Biomimetics, Reading, UK, pp 31-36

Cucchi V, Meredieu C, Stokes A, Berthier S, Bert D, Najar M, Denis A, Lastennet R (2004) Root anchorage of inner and edge trees in stands of Maritime pine (*Pinus pinaster* Ait.) growing in different soil podzolic conditions. Trees-Struct Funct 18:460-466

Danjon F, Barker DH, Drexhage M, Stokes A (2007) Using 3D plant root architecture in models of shallow slope stability. Ann Bot-London, in press

Danjon F, Fourcaud T, Bert D (2005) Root architecture and wind-firmness of mature *Pinus pinaster*. New Phytol 168:387-400

De Baets S, Poesen J, Knapen A, Barberá GG, Navarro JA (2007) Root characteristics of representative Mediterranean plant species and their erosion-reducing potential during concentrated runoff. Plant Soil 294:169-183

Dekker SC, Rietkerk M, Bierkens MFP (2007) Coupling microscale vegetation–soil water and macroscale vegetation–precipitation feedbacks in semiarid ecosystems. Global Change Biol 13:671-678

De Rooij GH (2000) Modeling fingered flow of water in soils owing to wetting front instability: a review. J Hydrol 231-232:277-294

DesRochers A, Lieffers VJ (2001) The coarse-root system of mature *Populus tremuloides* in declining stands in Alberta, Canada. J Veg Science 12:355-360

Dhakal AS, Sidle RC (2003) Long-term modelling of landslides for different forest management practices. Earth Surf Proc Land 28:853-868

Di Iorio A, Lasserre B, Scippa GS, Chiatante D (2005) Root system architecture of *Quercus pubescens* trees growing on different sloping conditions. Ann Bot-London 95:351-361

Di Iorio A, Lasserre B, Scippa GS, Chiatante D (2007) Pattern of secondary thickening in a *Quercus cerris* root system. Tree Physiol 27:407-412

Doorenbos J, Pruitt WO (1977) Crop water requirements. FAO, Rome

Duncan JM (1996) State of the art: limit equilibrium and finite element analysis of slopes. J Geotech Eng 122:577-596

Dupuy L, Fourcaud T, Stokes A (2005a) A numerical investigation into the influence of soil type and root architecture on tree anchorage. Plant Soil 278:119-134

Dupuy L, Fourcaud T, Stokes A (2005b) A numerical investigation into factors affecting the anchorage of roots in tension. Eur J Soil Sci 56:319-327

Dupuy L, Fourcaud T, Lac P, Stokes A (2007) A generic 3D finite element model of tree anchorage integrating soil mechanics and real root system architecture. Am J Bot 94: 1506–1514

Ekanayake JC, Phillips CJ (1999a) A model for determining thresholds for initiation shallow landslides under near-saturated conditions in the East Coast region, New Zealand. J Hydrol (NZ), 38, 1, 1-28

Ekanayake JC, Phillips CJ (1999b) A method for stability analysis of vegetated hillslopes: an energy approach. Can Geotech J 36:1172-1184

Ekanayake JC, Phillips CJ (2002) Slope stability thresholds for vegetated hillslopes: a composite model. Can Geotech J 39:849-862

Ekanayake JC, Marden M, Watson AJ, Rowan D (1997) Tree roots and slope stability: a comparison between *Pinus radiata* and kanuka. New Zeal J For Sci 27:216-233

Ennos AR (1989) The mechanics of anchorage in seedlings of sunflower, *Helianthus annuus* L. New Phytol 113:85-192

Ennos AR (1990) The anchorage of leek seedlings: the effect of root length and soil strength. Ann Bot-London 65:409-416

Ennos AR (1993) The scaling of root anchorage. J Theor Biol 161:61-75

Ennos AR (1994) The biomechanics of root anchorage. Biomimetics 2:129-137

Feddes RP, Kowalik P, Zaradny H (1978) Simulation of field water use and crop yield. Wageningen, Pudoc

Fourcaud T, Ji J-N, Zhang Z-Q, Stokes A (2007) Understanding the impact of root morphology on uprooting mechanisms: a modelling approach. Ann Bot-London, in press

Frydman S, Operstein V (2001) Numerical simulation of direct shear of root-reinforced soil. Ground Improv 5:41-48

Genet M, Stokes A, Salin F, Mickovski SB, Fourcaud T, Dumail J, van Beek LPH (2005) The influence of cellulose content on tensile strength in tree roots. Plant Soil 278:1-9

Genet M, Stokes A, Fourcaud T, Hu X, Lu Y (2006a) Soil fixation by tree roots: changes in root reinforcement parameters with age in *Cryptomeria japonica* D. Don. plantations. In: Marui H, Marutani T, Watanabe N, Kawabe H, Gonda Y, Kimura M, Ochiai H, Ogawa K,

Fiebiger G, Heumader J, Rudolf-Miklau F, Kienholz H, Mikos M. (eds) Interpraevent 2006: Disaster Mitigation of Debris Flows, Slope Failures and Landslides. September 25 – 27, 2006, Niigata, Japan. Universal Academy Press, Inc. Tokyo, Japan, ISBN 4-946443-98-3, pp 535-542

Genet M, Stokes A, Fourcaud T, Li M, Luo T (2006b) Effect of altitude on root mechanical and chemical properties of *Abies georgei* in Tibet. In: Salmen L (ed) Proceedings 5th Plant Biomechanics Conference, Sweden, 28 August – 1 September, 2006, pp 305-310

Gerrits AMJ, Savenije HHG, Hoffmann L, Pfister L (2006) Measuring forest floor interception in a beech forest in Luxembourg. Hydrol Earth Syst Sci Disc 3:2323-2341

Goodman AM, Ennos AR (1999) The effects of soil bulk density on the morphology and anchorage mechanics of the root systems of sunflower and maize. Ann Bot-London 83:293-302

Goss MJ (1987) The specific effect of roots on the regeneration of soil structure. In: Monnier G, Goss MJ (eds) Soil Compaction and Degeneration, Balkema, Boston, pp 145-155

Gray DH, Barker DH (2004) Root-soil mechanics and interactions. In: Bennett SJ, Collison AJC, Simon A (eds) Riparian vegetation and fluvial geomorphology: hydraulic, hydrologic and geotechnical interactions. Water Science and Application 8, American Geophysical Union, Washington, pp 125-139

Gray DH, Leiser AJ (1982) Biotechnical Slope Protection and Erosion Control. Van Nostrand Reinhold, New York

Gray DH, Megahan WF (1981) Forest Vegetation Removal and Slope Stability in the Idaho Batholith, United States Department of Agriculture Forest Service, Intermountain Forest and Range Experimental Station Research Paper, INT-271:1-23

Gray DH, Sotir RB (1996) Biotechnical and Soil Bioengineering Slope Stabilization: A Practical Guide for Erosion Control. Wiley & Sons, Inc., New York

Greenway DR (1987) Vegetation and slope stability. In: Anderson MG, Richards KS (eds) Slope Stability. Wiley, Chichester, pp 187-230

Greenwood JR (2006) Slip4ex - A program for routine slope stability analysis to include the effects of vegetation, reinforcement and hydrological changes. Geotech Geol Eng 24:449-465

Greenwood JR, Norris JE, Wint J (2004) Assessing the contribution of vegetation to slope stability. J Geotech Eng 157:199-208

Greenwood JR, Vickers AW, Morgan RPC, Coppin NJ, Norris JE (2001) Bioengineering - the Longham Wood Cutting field trial. CIRIA PR 81, London

Griffiths DV, Lane PA (1999) Slope stability analysis by finite elements. Geotechnique 49:387-403

Gruber F (1994) Morphology of coniferous trees: possible effects of soil acidification on the morphology of norway spruce and silver fir. In: Godbold D, Huttermann A (eds) Effects of Acid Rain on Forest Processes. Wiley & Sons, New York, pp 265-324

Gyssels G, Poesen J, Bochet E, Li Y (2005) Impact of plant roots on the resistance of soils to erosion by water: a review. Prog Phys Geog 29:189-217

Halter MR, Chanway CP (1993) Growth and root morphology of planted and naturally regenerated Douglas-fir and lodgepole pine. Ann Sci For 50:71-77

Hathaway RL, Penny D (1975) Root strength in some *Populus* and *Salix* clones. New Zeal J Bot 13:333-344

Hewlett JD, Hibbert AR (1967) Factors affecting the response of small watersheds to precipitation in humid areas. In: Supper WE, Lull HW (eds) International Symposium on Forest Hydrology. Pergamon, Oxford, pp 275-290

Hintikka V (1972) Wind-induced movements in forest trees. Comm Inst For Fenn 76:1-56

Hooghart J, Lablans W (1988) Van Penman naar Makkink. The Hague, CHO-TNO
Horton RE (1933) The role of infiltration in the hydrological cycle. Trans Am Geophysical Union 14:446-460
Horton RE (1945) Erosional development of streams and their drainage basins. Hydrophysical approach to quantitative morphology. Bull Geol Soc Am 56:275-370
Hsu LCY, Walker JCF, Butterfield BG, Jackson SL (2006) Compression wood does not form in the roots of *Pinus radiata.* IAWA J 27:45-54
Itasca (1993) FLAC – Fast Lagrangian Analysis of Continua. Minneapolis, USA
Jaffe MJ (1973) Thigmomorphogenesis: the response of plant growth and development to mechanical stimulation. Planta 114:143-157
Jaffe MJ, Telewski FW (1984) Thigmomorphogenesis: callose and ethylene in the hardening of mechanically stressed plants. In: Timmermann BN, Steelink C, Leowus FA (eds) Phytochemical Adaptations to Stress. Plenum Press, New York, pp 79-95
Jaffe MJ, Biro RL, Bridle K (1980) Thigmomorphogenesis: calibration of the parameters of the sensory function in beans. Physiol Plantarum 49:410-416
Jetten VG (1994) Modelling the effect of logging on the water balance of a tropical rainforest. A study in Guyana. PhD thesis, University Utrecht, Tropenbos Series 6. The Tropenbos Foundation, Wageningen, Netherlands
Johnson EA (1987) The relative importance of snow avalanche disturbance and thinning on canopy plant populations. Ecology 68:43-53
Kajimoto T, Daimaru H, Okamoto T, Otani T, Onodera H (2004) Effects of snow avalanche disturbance on regeneration of subalpine *Abies mariesii* forest, northern Japan. Arctic, Antarctic Alpine Res 36:436-445
Karrenberg S, Blaser S, Kollmann J, Speck T, Edwards PJ (2003) Root anchorage of saplings and cuttings of woody pioneer species in a riparian environment. Func Ecol 17:170-177
Khuder H (2007) L'architecture et les propriétés mécaniques des systèmes racinaires des arbres qui poussent en pente. PhD thesis, Université Bordeaux I, France
Khuder H, Danjon F, Stokes A, Fourcaud T (2006) Growth response and root architecture of Black locust seedlings growing on slopes and subjected to mechanical perturbation. In: Salmen L (ed) Proceedings 5th Plant Biomechanics Conference, Sweden, 28 August – 1 September, 2006, pp 299-304
Khuder H, Stokes A, Danjon F, Gouskou K (2007) Is it possible to manipulate root anchorage in young trees? Plant Soil 294:87-102
Kirkby MJ (1978) Hillslope Hydrology. Wiley, Chichester
Kokutse N, Fourcaud T, Kokou K, Neglo K, Lac P (2006) 3D numerical modelling and analysis of forest structure on hill slopes stability. In: Marui H, Marutani T, Watanabe N, Kawabe H, Gonda Y, Kimura M, Ochiai H, Ogawa K, Fiebiger G, Heumader J, Rudolf-Miklau F, Kienholz H, Mikos M (eds) Interpraevent 2006: Disaster Mitigation of Debris Flows, Slope Failures and Landslides. September 25 – 27, 2006, Niigata, Japan. Universal Academy Press, Inc. Tokyo, Japan, ISBN 4-946443-98-3, pp 561-567
Körner C (2003) Alpine Plant Life: Functional Plant Ecology of High Mountain Ecosystems. Springer, New York
Kostler JN, Bruckner E, Bibelriether H (1968) Die Wurzeln der Waldbaume Verlag Paul Parey Hamburg, Berlin
Kűlla T, Lõhmus K (1999) Influence of cultivation method on root grafting in Norway spruce (*Picea abies* (L.) Karst.). Plant Soil 217:91-100
Lewis GJ (1985) Root strength in relation to windblow. Forestry Commission Report on Forest Research. HMSO, London, pp 65-66

LI-COR (1992) LAI-2000 Plant canopy analyzer, Instruction Manual. LI-COR Inc., Lincoln, Nebraska, USA

Likens GE, Bormann FH, Pierce RS, Eaton JS, Johnson NM (1977) Biochemistry of a forested ecosystem. Springer Verlag, New York

Lindström A, Rune G (1999) Root deformation in containerised Scots pine plantations – effects on stability and stem straightness. Plant Soil 217:29-37

Lyford WH (1980) Development of the root system of northern red oak (*Quercus rubra* L.). Harvard Forest Paper 21

Makkink JF (1957) Testing the Penman formula by the use of lysimeters. J Inst Water Engrs 11:277-288

Makarova OV, Cofie P, Koolen AJ (1998) Axial stress-strain relationships of fine roots of Beech and Larch in loading to failure and in cyclic loading. Soil Till Res 45:175-187

Malet J-P, van Asch ThWJ, van Beek LPH, Maquaire O (2003) Apport des models hydrologiques spatialises a la simulation numerique de glissements de terrain. Impact pour la gestion du risque., SIRNAT – Les journees pour la prevention des risques. BRGM reports. BRGM, Orleans

Marler TE, Discekici HM (1997) Root development of 'red lady' papaya plants growing on a hillside. Plant Soil 195:37-42.

Marshall TJ, Holmes JW (1988) Soil Physics. Cambridge University Press, Cambridge

Mattia C, Bischetti GB, Gentile F (2005) Biotechnical characteristics of root systems of typical Mediterranean species. Plant Soil 278:23-32

Mickovski SB, Ennos AR (2003) The effect of unidirectional stem flexing on shoot and root morphology and architecture in young *Pinus sylvestris* trees. Can J For Res 33:2022-2029

Mickovski SB, Van Beek LPH, Salin F (2005) Uprooting resistance of vetiver grass (*Vetiveria zizanioides*). Plant Soil 278:33-41

Mickovski SB, Bengough AB, Bransby MF, Davies MCR, Hallett PD, Sonnenberg R (2007) Material stiffness, branching pattern and soil matric potential affect the pullout resistance of model root systems. Eur J Soil Sci, in press, doi: 10.1111/j.1365-2389.2007.00953

Nicoll BC, Berthier S, Achim A, Gouskou K, Danjon F, van Beek LPH (2006) The architecture of *Picea sitchensis* structural root systems on horizontal and sloping terrain. Trees-Struct Func 20:701-712

Nicoll BC, Ray D (1996) Adaptive growth of tree root systems in response to wind action and site conditions. Tree Physiol 16:891-898

Niklas KJ (1999) Variations of the mechanical properties of *Acer saccharum* roots. J Exp Bot 50:193-200.

Nilaweera NS, Nutalaya P (1999) Role of tree roots in slope stabilisation. Bull Eng Geol Env 57:337-342

Nörr R (2003) Planting – a risk for the stability of forest stands? In: Ruck A, Kottmeier C, Mattheck C, Quine C, Wilhelm G (eds) Wind Effects on Trees, International Conference, 16-18 Sept. 2003, Karlsruhe, Germany, pp 281-288

Norris JE (2005a) Root reinforcement by hawthorn and oak roots on a highway cut-slope in Southern England. Plant Soil 278:43-53

Norris JE (2005b) Root mechanics applied to slope stability. PhD thesis, Nottingham Trent University, Nottingham, UK

Norris JE, Greenwood JR (2003) Root reinforcement on unstable slopes in Northern Greece and Central Italy. International Conference on Problematic Soils, July 2003, Nottingham Trent University, Nottingham, UK, pp 411-418

Operstein V, Frydman S (2002) The stability of soil slopes stabilised with vegetation. Ground Improv 6:163-168
O'Loughlin CL, Watson A (1979) Root-wood strength deterioration in Radiata Pine after clearfalling. New Zeal J Forest Sci 9:284-93
Parlange JY, Smith RE (1976) Ponding time for variable rainfall rates. Can J Soil Sci 56:212-223
Parr A, Cameron AD (2004) Effects of tree selection on strength properties and distribution of structural roots of clonal Sitka spruce. For Ecol Manage 195:97-106
Patel RN (1964) On the occurrence of gelatinous fibres with special reference to root wood. J Inst Wood Sci 12:67-80
Peltola H (2006) Mechanical stability of trees under static loads. Am J Bot 93:1501-1511
Penman HL (1948) Natural evaporation from open water, bare soil and grass. Proc Royal Soc A193:120-145
Perry J (1989) A survey of slope condition on motorway earthworks in England and Wales. RR199, Transport and Road Research Laboratory, Crowthorne.
Perry J, Pedley M, Brady K (2003a) Infrastructure cuttings – condition appraisal and remedial treatment. C591, CIRIA, London
Perry J, Pedley M, Reid M (2003b) Infrastructure embankments – condition appraisal and remedial treatment. C592, 2nd edition, CIRIA, London
Pollen N (2006) Temporal and spatial variability in the root-reinforcement of streambanks: Accounting for soil shear strength and moisture. Catena 69:197-205
Pollen N, Simon A (2005) Estimating the mechanical effects of riparian vegetation on stream bank stability using a fiber bundle model, Water Resour Res 41:W07025, doi:10.1029/2004WR003801
Preisig CL, Carlson WC, Promnitz LC (1979) Comparative root system morphologies of seeded-in-place, bare-root and containerised Douglas-fir seedlings after out-planting. Can J Forest Res 9:399-405
Priestly CHB, Taylor RJ (1972) On the assessment of surface heat flux and evaporation using large-scale parameters. Mon Weather Rev 100:81-92
Rietkerk M, Dekker SC, de Ruiter PC, van de Koppel J (2004) Self-organized patchiness and catastrophic shifts in ecosystems. Science 305:1926-1929
Reubens B, Poesen J, Danjon F, Geudens G, Muys B (2007) The role of fine and coarse roots in shallow slope stability and soil erosion control with a focus on root system architecture: a review. Trees-Struct Func 4:385-402
Ridolfi L, D'Odorico P, Porporato A, Rodriquez-Iturbe I (2003) Stochastic soil moisture dynamics along a hillslope. J Hydrol 272:264-275
Riedl H (1937) Bau und leistungen des wurzelholzes. Jahrbücher für Wissenschaftliche Botanik. Leipzig, Germany, Verlag von Gebrüder Borntrager, pp 1-75
Rune G (2003) Slits in container wall improve root structure and stem straightness of outplanted Scots pine seedlings. Silva Fenn 37:333-342
Sakals ME, Sidle RC (2004) A spatial and temporal model of root cohesion in forest soils. Can J For Res 34:950-958
Schiechtl HM (1980) Bioengineering for Land Reclamation and Conservation. University of Alberta Press, Edmonton, Alberta, Canada
Schiechtl HM, Stern R (1996) Water Bioengineering Techniques: For Watercourse, Bank and Shoreline Protection. Blackwell Publishing Limited
Schmidt KM, Roering JJ, Stock JD, Dietrich WE, Montgomery DR, Schaub T (2001) The variability of root cohesion as an influence on shallow landslide susceptibility in the Oregon Coast Range. Can Geotech J 38:995-1024

Scippa GS, Di Michele M, Di Iorio A, Costa A, Lasserre B, Chiatante D (2006) Root response to slope: anchorage and gene factors involved. Ann Bot-London 97:857-866

Shrestha MB, Horiuchi M, Yamadera Y, Miyazaki T (2000) A study on the adaptability mechanism of tree roots on steep slopes. In: Stokes A (ed) The Supporting Roots of Trees and Woody Plants: Form, Function and Physiology. Developments in Plant and Soil Sciences, vol. 87. Kluwer Academic Publishers, Dordrecht, Netherlands, pp 51-57

Sidle RC (1992) A theoretical model of the effects of timber harvesting on slope stability. Water Res Res 28:1897-1910

Simon A, Collison AJC (2002) Quantifying the mechanical and hydrologic effects of riparian vegetation on streambank stability. Earth Surf Proc Land 27:527-546

Soethe N, Lehmann J, Engels C (2006) Root morphology and anchorage of six native tree species from a tropical montane forest and an elfin forest in Ecuador. Plant Soil 279: 173-185

Steele DP, MacNeil DJ, McMahon W, Barker DH (2004) The use of live willow poles for stabilising highway slopes. TRL Report 619, Crowthorne, TRL Limited

Stone EL, Kalisz PJ (1991) On the maximum extent of tree roots. Forest Ecol Manag 46: 59-102

Stokes A (1999) Strain distribution during anchorage failure in root systems of Maritime pine (*Pinus pinaster* Ait.) at different ages and tree growth response to wind-induced root movement. Plant Soil 217:17-27

Stokes A (2002) The biomechanics of tree root anchorage. In: Waisel Y, Eshel A, Kafkaki U (eds) Plant Roots – The Hidden Half. Plenum Publishing, New York, pp 175-186

Stokes A, Guitard DG (1997) Tree root response to mechanical stress. In: Altman A, Waisel Y (eds) The Biology of Root Formation and Development. Plenum Publishing, New York, pp 227-236

Stokes A, Mattheck C (1996) Variation of wood strength in tree roots. J Exp Bot 47:693-699

Stokes A, Fitter AH, Coutts MP (1995) Responses of young trees to wind and shading: effects on root architecture. J Exp Bot 46:1139-1146

Stokes A, Berthier S, Sacriste S, Martin F (1998) Variations in maturation strains and root shape in root systems of Maritime pine (*Pinus pinaster* Ait.). Trees-Struct Func 12:334-339.

Stokes A, Nicoll BC, Coutts MP, Fitter AH (1997) Responses of young Sitka spruce clones to mechanical perturbation and nutrition: effects on biomass allocation, root development, and resistance to bending. Can J Forest Res 27:1049-1057

Stokes A, Abd.Ghani M, Salin F, Danjon F, Jeannin H, Berthier S, Kokutse AD, Frochot H (2007a) Root morphology and strain distribution during tree failure on mountain slopes. In: Stokes A, Spanos I, Norris JE, Cammeraat LH (eds) Eco- and Ground Bio-Engineering: The Use of Vegetation to Improve Slope Stability. Developments in Plant and Soil Sciences Volume 103, Springer Publishers, Dordrecht, pp 165-173

Stokes A, Lucas A, Jouneau L (2007b) Plant biomechanical strategies in response to frequent disturbance: uprooting of *Phyllostachys nidularia* (Poaceae) growing on landslide prone slopes in Sichuan, China. Am J Bot 94:1129-1136

Stokes A, Salin F, Kokutse AD, Berthier S, Jeannin H, Mochan S, Kokutse N, Dorren L, Abd.Ghani M, Fourcaud T (2005) Mechanical resistance of different tree species to rockfall in the French Alps. Plant Soil 278:107-117

Storey PJ (2002) The conservation and improvement of sloping land: a manual of soil and water conservation and soil improvement on sloping land. Volume 1. Science Publishers, Enfield, New Hampshire, USA

Styczen ME, Morgan RPC (1995) Engineering properties of vegetation. In: Morgan RPC, Rickson RJ (eds) Slope Stabilisation and Erosion Control: a Bioengineering Approach. McGraw Hill, pp 5-58

Telewski FW (1995) Wind-induced physiological and developmental responses in trees. In: Coutts MP and Grace J (eds) Wind and Tree. Cambridge University Press, Cambridge, pp 237-263

Telewski FW (2006) A unified hypothesis of mechanoperception in plants. Am J Bot 93:1466-1476

Terwilliger VJ, Waldron LJ (1991) Effects of root reinforcement on soil-slip patterns in the transverse ranges of southern California. Geol Soc Am Bull 103:775-785

Timell TE (1986) Compression Wood in Gymnosperms. Springer-Verlag, Berlin

Turmanina VI (1965) On the strength of tree roots. Bull Moscow Soc Nat Biol Sect 70:36-45

Valentin C, d'Herbers JM, Poesen J (1999) Soil and water components of banded vegetation patterns. Catena 37:1-24

Van Asch ThWJ, van Dijck SJE, Hendriks MR (2001) The role of overland flow and subsurface flow on the spatial distribution of soil moisture in the topsoil. Hydrol Process 15:2325-2340

Van Beek LPH (2002) Assessment of the influence of changes in land use and climate on landslide activity in a Mediterranean environment. Netherlands Geographical Studies 294, KNAG, Utrecht

Van Beek LPH, van Asch ThWJ (1998) A combined conceptual model for the effects of fissure-induced infiltration on slope stability. In: Hergarten S, Neugebauer HJ (eds) Process Modelling and Landform Evolution. Springer Verlag, Berlin, pp 147-169

Van Beek LPH, Cammeraat LH (2007) Infiltration and soil water redistribution under different types of land cover after abandonment: field observations from broadscale rainfall experiments. Hydrol Process, in press

Van Beek LPH, Wint J, Cammeraat LH, Edwards JP (2005) Observation and simulation of root reinforcement on abandoned Mediterranean slopes. Plant Soil 278:55-74

Waldron LJ (1977) The shear resistance of root permeated homogenous and stratified soil. J Soil Sci Soc Am 41:843-849

Waldron LJ, Dakessian S (1981) Soil reinforcement by roots: calculation of increased soil shear resistance from root properties. Soil Sci 132:427-435

Waldron LJ, Dakessian S (1982) Effect of grass, legume, and tree roots on soil shearing resistance. J Soil Sci Soc Am 46:894-899

Wang WL, Yen BC (1974) Soil arching in slopes. J Geotech Eng Div ASCE, 100, GT1:61-78

Wasterlund I (1989) Strength components in the forest floor restricting maximum tolerable machine forces. J Terramech 26:177-182

Watson AJ, Marden M, Rowan D (1995) Tree species performance and slope stability. In: Barker DH (ed) Vegetation and Slope Stabilisation, Protection and Ecology, Thomas Telford Press, London, pp 161-171

Willatt ST, Sulistyaningsih N (1990) Effect of plant roots on soil strength. Soil Till Res 16:329-336

Wu TH (1976) Investigation of landslides on Prince of Wales Island. Geotechnical Engineering Report 5, Civil Engineering Department, Ohio State University, Columbus, Ohio, USA

Wu TH, McKinnell III WP, Swanston DN (1979) Strength of tree roots and landslides on Prince of Wales Island, Alaska. Can Geotech J 16:19-33

Wu TH, Watson AJ, El-Khouly MA (2004) Soil-root interaction and slope stability. In: Barker DH, Watson AJ, Sombatpanit B, Northcut B, Magliano AR (eds) Ground and Water Bioengineering for Erosion Control and Slope Stabilization. Science Publishers Inc. USA, pp 183-192

Zhou Y, Watts D, Cheng X, Li Y, Luo H, Xiu Q (1997) The traction effect of lateral roots of *Pinus yunnanensis* on soil reinforcement: a direct *in situ* test. Plant Soil 190:77-86

Ziemer RR (1981) Roots and shallow stability of forested slopes. Int Ass Hydrol Sci 132:343-361

Zienkiewicz OC, Humphreson C, Lewis RW (1975) Associated and non-associated visco-plasticity and plasticity in soil mechanics. Geotechnique 25:671-689

Zienkiewicz OC, Taylor RL (1998) The Finite Element Method, 4th edn, vol. 2. Solid and Fluid Mechanics, Dynamics and Non-linearity. McGraw-Hill, New York

Zinke P (1967) Forest interception studies in the United States. In: Sopper WE, Lull HW (eds) International Symposium on Forest Hydrology, Proc. National Science Foundation Adv. Science Seminar, Pennsylvania State University, University Park, 29 August–10 September 1965. Pergamon Press, Oxford, UK, pp 137-161

Chapter 5

HAZARD ASSESSMENT OF VEGETATED SLOPES

Joanne E. Norris[1,2], John R. Greenwood[2], Alexis Achim[3], Barry A. Gardiner[4], Bruce C. Nicoll[4], Erik Cammeraat[5], Slobodan B. Mickovski[6]

[1] Halcrow Group Limited, Endeavour House, Forder Way, Cygnet Park, Hampton, Peterborough, PE7 8GX, U.K., [2] School of Architecture, Design and Built Environment, Nottingham Trent University, Burton Street, Nottingham, NG1 4BU, U.K., [3] Faculté de Foresterie et de Géomatique, Université Laval, Québec, G1K 7P4, Canada, [4] Forest Research, Northern Research Station, Roslin, Midlothian, EH25 9SY, U.K., [5] IBED-Physical Geography, University of Amsterdam, Nieuwe Achtergracht 166, 1018 WV Amsterdam, The Netherlands, [6] Jacobs UK Ltd., Glasgow, G2 7HX, U.K.

Abstract: *The hazard assessment of vegetated slopes are reviewed and discussed in terms of the stability of the slope both with and without vegetation, soil erosion and the stability of the vegetated slope from windthrow and snow loading. Slope stability can be determined by using either limit equilibrium or finite element stability analysis methods. The limit equilibrium methods are extended to incorporate the vegetation parameters that are important for the stability of a vegetated slope. The factors that contribute to soil erosion are reviewed and the techniques for assessing and measuring the rate of soil erosion are presented. The assessment of windthrow hazards are comprehensively discussed and a mechanistic model called ForestGALES is introduced which has flexibility for testing many different forest management scenarios. The hazards presented by snow loading on forested slopes are briefly reviewed.*

Key words: hazard assessment, slope stability, soil erosion, vegetated slopes, windthrow.

1. INTRODUCTION

Hazards may be defined as sources of potential harm resulting from natural processes (natural hazards) or human activity (man-made hazards). The risk of a hazardous event occurring can be assessed in terms of the

J.E. Norris et al. (eds.), Slope stability and erosion control: Ecotechnological solutions, 119–166.

probability and possible impact of the event. In this chapter, a limited number of natural and man-made hazards and their determination is discussed, and related to various processes on slopes. The following hazards are elaborated in detail:

- Slope (in)stability (Sections 1.1 and 1.2)
- Soil erosion (Section 2)
- Stability of vegetation on slopes from windthrow and snow hazards (Section 3)

and general techniques to assess hazards, i.e.,

- Mapping inventory techniques, both in the field and using aerial photographs/remote sensing techniques
- Geographical Information Systems (GIS) techniques
- Numerical modelling
- Decision support systems

Before starting the actual assessment it is necessary to make some general remarks on the assessment related to slope characteristics, soil materials and vegetation.

Initially, a simple inventory should be carried out, focussing in particular on the presence of:

- Signs of mass wasting, slope angle and sudden slope breaks, susceptible geological and soil materials, adverse hydrological conditions and topographical surfaces, e.g., areas showing signs of mass wasting may include sudden slope breaks and materials with adverse soil mechanical properties, e.g., certain clay rich materials. Unfavourable hydraulic conditions may also exist, e.g., spring zones and badly drained areas.
- Erosion processes and vegetation damage from the past.

Areas showing signs of soil erosion may be indicated by partial or absent vegetation cover, truncated soil profiles, erodibility of soil material as well as land use practices and soils with impervious layers close to the surface.

Areas where *vegetation* is or has historically been known to be damaged by several processes, e.g., forest fires, storms, diseases or insect invasions, are also susceptible.

Artificial slopes need special attention (both existing and designed). Two main types can be distinguished:

1. *piled up materials*. Artifical slopes consisting of loosely piled materials often show a lack of cohesion and internal strength, making them very sensitive to slope failure or rill and gully erosion.
2. *consolidated materials*. Artificial slopes consisting of compacted and consolidated clays are prone to slope failure if design errors have been incurred, related to the over-steepening of slopes and tension release after cutting the slope.

Following the initial assessment, in which a Slope Decision Support System (Mickovski et al. 2005; Mickovski and van Beek 2006) might be of help, more detailed methods can be used which are discussed in the following sections.

Risk assessment of the hazards described here is only partly addressed in this chapter. For further description of this, the reader is referred to standard textbooks on hazard risk assessment (Glade et al. 2005).

1.1 Slope stability assessment

When assessing the stability of a slope, either vegetated or non vegetated, certain information is required on the topography, site layout, geology, soil and groundwater conditions that may be present or are likely to be encountered. Slopes generally fail on either geologically weak points in rock slopes or on shear planes in soil slopes. The conditions along a potential failure surface must, therefore, be defined in terms of:

- Normal stress acting on the failure surface
- Pore water pressure
- Shear strength of the material intersected by the failure surface
- Pull out forces generated by soil reinforcements or anchors.

The stability of slopes may conveniently be analysed by limit equilibrium methods, e.g., Duncan and Wright (2005). Limit equilibrium analysis requires information about the strength of the soil, but not its stress-strain behaviour. Slope movements are usually analysed by finite-element methods i.e., finite element software programs such as PLAXIS (http://www.plaxis.nl/). For these methods, characteristic stress-strain behaviour is required.

1.1.1 Slope stability analysis by limit equilibrium methods

In limit equilibrium techniques, e.g., Bishop (1955) and Fellenius (1936), the stability of a possible slip surface is assessed by comparing the gravitational disturbing forces with the available shearing resistance (shear strength) of the ground along the slip surface (Figure 5.1). For stability, disturbing forces acting along all potential slip planes must be less than the resisting forces that can be mobilised along them. The disturbing forces are due to the self weight of the material lying above the failure surface and to any external loads. Resisting forces are generated by the strength of the soil and by the pull out forces generated by soil reinforcement (for instance, the roots of vegetation). For stability to be maintained the available shear strength must exceed the disturbing forces.

The Factor of Safety (FOS) against failure is expressed by:

$$FOS = \frac{shear\ resistance}{shear\ force\ required\ for\ equilibrium} = \frac{restoring\ force}{disturbing\ force} \quad (1)$$

The FOS is generally expressed in terms of moment equilibrium, where the FOS for a stable slope will be greater or equal to 1.

For a circular slip surface, FOS is expressed in terms of moment equilibrium (FOS_m) with the lever arm (radius R) cancelling from the numerator and denominator of the equation.

For non-circular slip surfaces, FOS may be assumed to be expressed in terms of pseudo-moment equilibrium (with a changing value of R which is assumed to cancel from the numerator and denominator).

The FOS might also be expressed in terms of horizontal force equilibrium (FOS_f) for compatibility with retaining structure design.

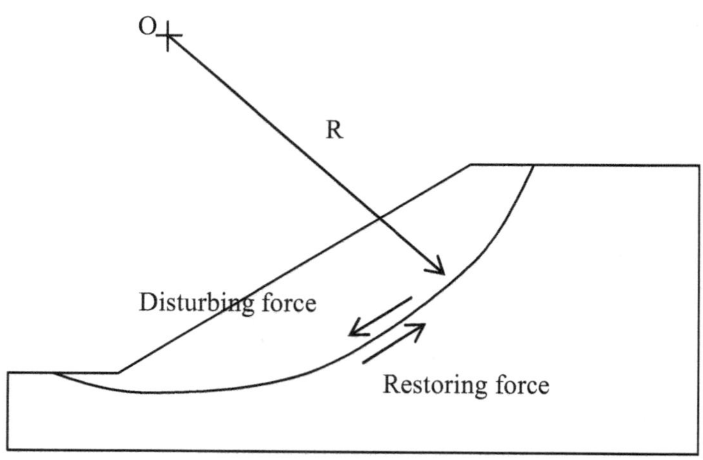

Figure 5-1. Forces acting on a circular slip plane. O is the centre of the slip circle, R is the radius of the slip circle or lever arm.

Method of Slices

The FOS for a slope is normally derived by the method of slices (Duncan and Wright 2005; Greenwood 2006). This method uses the friction block acting on an inclined plane as the basis for stability analysis. A block or slice of soil of unit width, above a potential slip surface, has the same friction principles applied to control stability but now there is the added effect of soil cohesion and water pressure which will govern the effective stresses.

To determine the FOS by the method of slices, a circular slip surface with radius R is assumed. The soil mass above the arc is divided into a number of vertical slices of width b and varying height h (Figure 5.2). The base of each slice is assumed to be a straight line inclined at an angle α to the horizontal and with a length l (Figure 5.2). The slope is divided into slices for analysis purposes only. It is assumed that all slices rotate around the centre of the circle O as a whole body. This implies that forces must act between the slices, termed interslice forces.

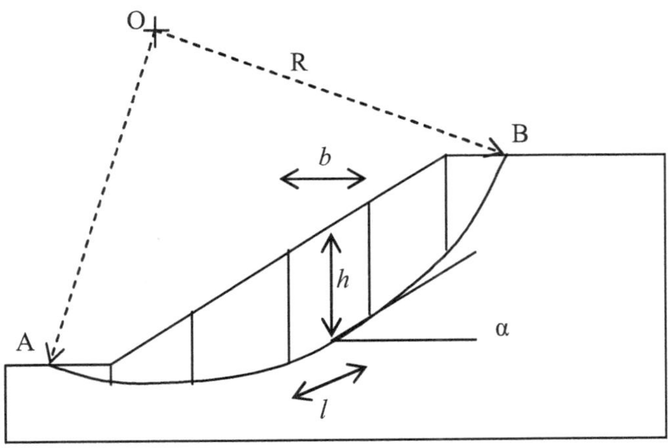

Figure 5-2. Method of slices. A circular slip surface of radius R, has centre O and intersection points at the ground surface of A and B. The soil mass above the slip surface is divided into a number of vertical slices of width b and varying height h. The base of each slice is assumed to be a straight line inclined at an angle α to the horizontal and with a length l.

The forces acting on a slice (Figure 5.3) are:

- The total weight of the slice, $W = \gamma bh$ where γ is the bulk unit weight of the soil.
- The weight of each slice induces a shear force parallel to its base, $S = W\sin\alpha$.

- The total normal force on the base, $N = \sigma l$.
- The total normal force is obtained from total normal stress, i.e., the effective normal force $N' = \sigma' l$ and the water force $U = ul$ where u is the pore water pressure.
- The shear force τl.
- The interslice forces, represented as total normal forces E_1 and E_2 and tangential shear forces X_1 and X_2.

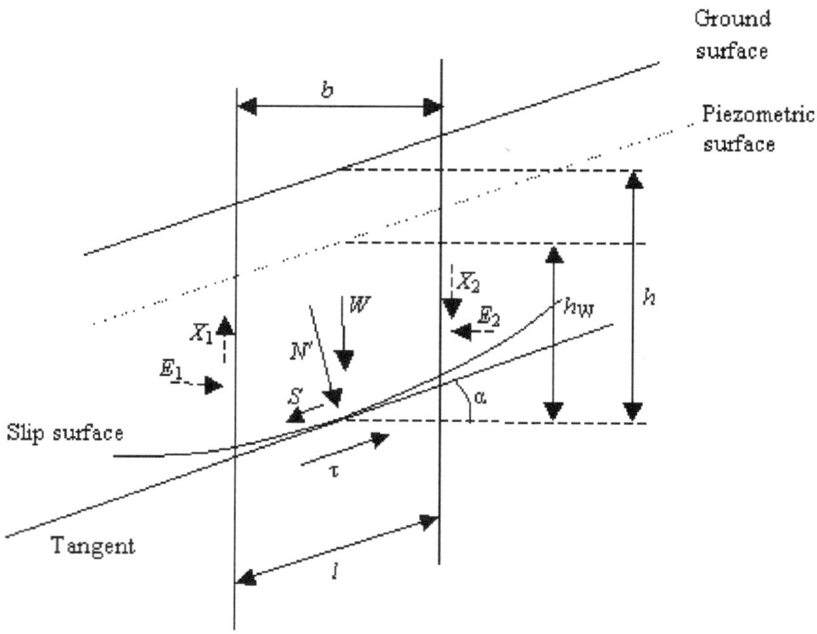

Legend:

W	Weight of slice
h	Average height of slice
h_W	Head of water above slip surface
α	Angle of base of slice
l	Length of slip surface
b	Width of slice ($b = l\cos\alpha$)
N'	Effective normal force on slip surface
u	Water pressure = $\gamma_W h_W$
τ	Shear strength
X_1, X_2, E_1, E_2	Interslice forces

Figure 5-3. Forces acting on a slice.

For each slice, FOS is given by (from Figure 5.3):

$$FOS = \frac{\tau l}{W sin\alpha} \tag{2}$$

By applying the Mohr-Coulomb strength relationship, i.e., $\tau = c' + \sigma'_n \tan \phi'$ where τ = available shear stress, c' = effective cohesion, σ'_n = effective normal stress on the shear plane and ϕ' = effective angle of friction at the slip surface. Equation [2] can now be written as:

$$FOS = \frac{c'l + N' \tan \phi'}{W \sin \alpha} \tag{3}$$

where $N' = \sigma'_n l$.

The effects of the single slice may now be added to the adjacent slices to give the overall FOS for the slip surface.

$$FOS = \frac{\sum c'l + N' \tan \phi'}{\sum W \sin \alpha} \tag{4}$$

The value N' in Equation [4] may be determined by resolving forces, where $N' = W \cos\alpha - ul + (X_2 - X_1)\cos\alpha - (E_2 - E_1)\sin\alpha$, i.e.,

$$FOS = \frac{\sum (c'l + (W \cos\alpha - ul) \tan \phi' + [(X_2 - X_1)\cos\alpha - (E_2 - E_1)\sin\alpha] \tan \phi')}{\sum W \sin \alpha} \tag{5}$$

However, to solve Equation [5] assumptions must be made regarding the interslice forces. Table 5-1 shows the solutions to the interslice force assumptions made by Fellenius (1936), Bishop (1955), Janbu (1973) and Greenwood (1987).

NB., The FOS value must be determined for the surface that is likely to fail, i.e., the critical slip surface. It is therefore necessary to perform calculations for a considerable number of possible slip surfaces in order to determine the location of the critical slip surface.

Table 5-1. Solutions and assumptions to the Factor of Safety equation.

Method	**FOS Equation**	**Assumptions**
Fellenius	$\frac{\Sigma[c'l + (W\cos\alpha - ul)\tan\phi']}{\Sigma W\sin\alpha}$	Water surface is parallel to the slip surface, i.e., $(X_2 - X_1)\cos\alpha - (E_2 - E_1)\sin\alpha = 0$. NB. Considerable errors occur when steep base angles to the slice are combined with high water pressures (Turnbull and Hvorslev 1967; Greenwood 1983).
Bishop	$\frac{\Sigma\left[\frac{(c'b + (W - ub)\tan\phi')\sec\alpha}{(1 + (1/FOS_m)\tan\phi'\tan\alpha)}\right]}{\Sigma W\sin\alpha}$	Tangential interslice forces are equal and opposite ($X_1 = X_2$) and the normal interslice forces are not equal ($E1 \neq E2$). NB. The value of FOS occurs on both sides of the expression, therefore an estimated value for FOS must be chosen on the right hand side to obtain a value of FOS on the left hand side. By successive iteration convergence on the true value of FOS is obtained.
Janbu	$\frac{\Sigma\left[\frac{(c'b + (W - ub)\tan\phi')\sec\alpha}{(1 + (1/FOS_f)\tan\phi'\tan\alpha)\cos\alpha}\right]}{\Sigma W\tan\alpha} x f_0$	Identical to Bishop except that the equation is expressed in terms of horizontal force equilibrium and a compensation multiplying factor is introduced (typically $f_0 = 1.05$).
Greenwood General	$\frac{\Sigma[c'l + (W\cos\alpha - ul - (U_2 - U_1)\sin\alpha)\tan\phi']}{\Sigma W\sin\alpha}$	Effective interslice forces analysed and water forces, U_1 and U_2, on the sides of the slice are taken into account, i.e., $(X'_2 - X'_1)\cos\alpha - (E'_2 - E'_1)\sin\alpha = 0$.

Greenwood General (with K)	$\dfrac{\sum(c'l + [W\cos\alpha - ul - (U_2 - U_1)\sin\alpha + K\tan\alpha(W - ub)\sin\alpha]\tan\phi')}{\sum W\sin\alpha}$	Inclusion of coefficient of horizontal earth pressure, K, influences position of critical slip surface (particularly in over-consolidated soils).

Horizontal force equilibrium

It is sometimes convenient to express the FOS in terms of horizontal force equilibrium (FOS_f), e.g., for slips involving a significant near horizontal movement or to relate to retaining wall design. The equivalent horizontal forces are determined for each slice of the analysis simply by dividing the numerator and denominator of the stability equation by $\cos\alpha$. The Greenwood General (Greenwood 1989, 1990; Morrison and Greenwood 1989), and Fellenius equations may all be converted to horizontal force equilibrium in the same way as the Bishop equation converts to the Janbu equation.

Confidence in the Factor of Safety

An acceptable FOS for a particular slope requires sound engineering judgment due to the multiple factors which must be considered. A qualified geotechnical engineer must be consulted in all cases. A FOS for a slope can only be determined when there is an appropriate method of analysis; flow slides and erosion are not readily analysed by these methods.

For each slope, two factors should be considered: (1) the consequences of failure occurring and (2) the confidence in the information available. When there is a risk to life and adjacent structures a higher FOS would be normally be chosen. A lower FOS is chosen when instabilities do not affect lives or structures. The FOS is very dependent on the complexity of the ground conditions, the quality of the data obtained from the site investigation and the certainty of the design parameters.

The FOS selected is very dependent on the confidence in the parameters selected for the analysis. For a slope on the point of failure a remedial action that increased the FOS calculated by back analysis[1] by say 5% from 1.00 to 1.05 would provide greater confidence than a calculated value of 1.05 based on estimated parameters. It should be noted that in accordance with recent European standards (BS EN1997-2 2007) 'partial' safety factors are now applied to individual parameters of stability equations to reflect the level of confidence in that parameter.

[1] A failed slope is considered to have a FOS of unity (1.0) at the time of failure. Using this knowledge and an appropriate method of analysis, a model of the slope at failure can be developed. The process by which the failure conditions are determined and the failure model is established is termed back analysis or back calculation (Duncan and Wright 2005).

UK recommendations for cuttings, natural slopes and embankments are for FOS between 1.3 and 1.4 for first time slides and a FOS of 1.2 for slides with pre-existing slip surfaces (BS6031 1981).

1.2 Vegetation factors in slope stability

In this chapter, we are primarily concerned with the stability of vegetated slopes or slopes that have the potential to be vegetated. The influences of vegetation on a slope and the modification of the basic stability equation to include the effects of vegetation are therefore discussed.

Figure 5.4 shows the additional parameters that need to be considered when incorporating vegetation into the stability analysis. Each additional parameter is explained in the following sections and values are suggested for different vegetation types for input in the stability analysis. The parameters are further discussed in Coppin and Richards (1990) and Greenwood et al. (2004).

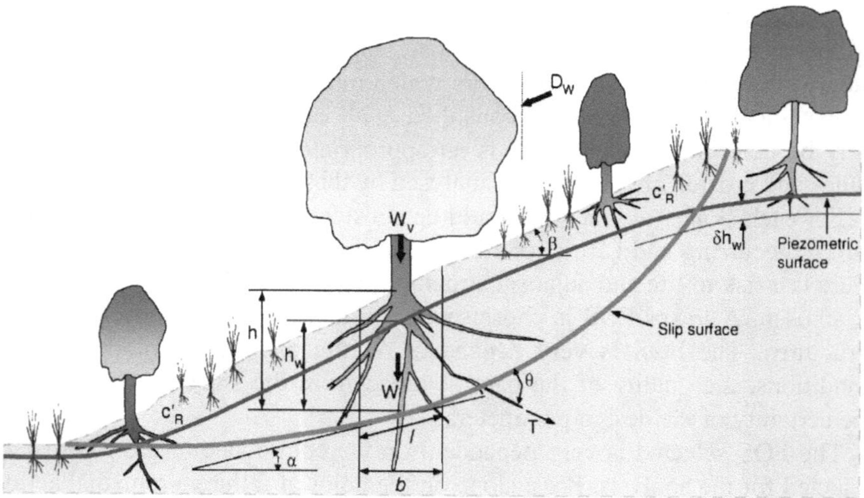

Figure 5-4. Forces exerted on a slope by vegetation (after Greenwood et al. 2004). Parameters: α – angle of slip surface; β – slope angle; c'_R – enhanced cohesion due to fine roots; D_w – wind force; b – width of slice; l – length of slice; h_z – height of slice above slip surface; h_w – height of phreatic surface above slip surface; δh_w – change in phreatic surface due to uptake of water by vegetation; W – total weight of soil slice; W_v – surcharge of vegetation; T – tensile force of roots acting on slip surface; θ – angle of roots to slip surface.

Enhanced cohesion, c'_R

The concept of effective cohesion in soils has received considerable attention with some researchers advocating that no true cohesion exists in clay soils (Schofield 1998, 1999; Goodman 1999). However, back analysis of slope failures has generally indicated an operational effective shear strength which is conveniently represented by a small cohesion intercept in the order of $c' = 1\text{–}2$ kPa. The actual value of c'_R input into the slope stability analysis can have considerable influence on the calculated FOS. Values of c'_R have been measured by researchers often based on direct *in situ* shear tests, back analysis or from root density and vertical root model equations (Table 5-2). Values vary from 1–25 kPa depending on the type of soil and vegetation. Tests carried out by Schmidt et al. (2001) show that lateral root cohesion ranges from 6.8–23.2 kPa for industrial forests with understory and deciduous vegetation, 25.6–93.4 kPa for natural forests dominated by coniferous vegetation and ≤10 kPa in clear-cut areas from the Oregon Coast Range (Table 5-3).

In situ shear apparatus (Figure 5.5) can be readily manufactured in the workshop and with a team of volunteers, a number of shear tests can be carried out in a day (Norris and Greenwood 2003; Norris 2005a, b; van Beek et al. 2005). Field tests will tend to give an indicative undrained strength increase due to the presence of fine roots but, for clay soils, the true effective parameters are more accurately obtained by back analysis or more sophisticated effective stress laboratory testing.

The use of enhanced c' values is appropriate for grassed areas or areas of uniform vegetation where fine root distribution with depth is consistent and easily defined. In general, the reliable benefit of an enhanced c' value will be limited to shallow depths.

Figure 5-5. Set up of *in situ* shear apparatus (Photo: J.E. Norris).

Table 5-2. Typical values for increases in soil cohesion (c'_R) due to roots (updated from Norris and Greenwood 2006).

Source	**Vegetation, soil type and location**	**Root cohesion c'_R (kPa)**
	Grass and Shrubs	
Wu[3] (1984a)	Sphagnum moss (*Sphagnum cymbifolium* L.), Alaska, USA	3.5 – 7.0
Barker[2] (1987)	Boulder clay fill (dam embankment) under grass in concrete block reinforced cellular spillways, Jackhouse Reservoir, UK	3.0 – 5.0
Buchanan and Savigny[1] (1990)	Understorey vegetation (*Alnus*, *Tsuga*, *Carex*, *Polystichum*), glacial till soils, Washington, USA	1.6 – 2.1
Gray[5] (1995)	Reed fiber (*Phragmites communis* Trin.) in uniform sands, laboratory	40.7
Tobias[2] (1995)	*Alopecurus geniculatus* L., forage meadow, Zurich, Switzerland	9.0
Tobias[2] (1995)	*Agrostis stolonifera* L., forage meadow, Zurich, Switzerland	4.8 – 5.2
Tobias[2] (1995)	Mixed pioneer grasses (*Festuca pratensis* Huds., *Festuca rubra* L., *Poa pratensis* L.), alpine, Reschenpass, Switzerland	13.4
Tobias[2] (1995)	*Poa pratensis* L. (monoculture), Switzerland	7.5
Tobias[2] (1995)	Mixed grasses (*Lolium multiflorum* Lam., *Agrostis stolonifera* L., *Poa annua* L.), forage meadow, Zurich, Switzerland	–0.6 – 2.9
Cazzuffi et al.[5] (2006)	Elygrass (*Elytrigia elongata L.*) Eragrass (*Eragrostis curvala* Nees) Pangrass (*Panicum virgatum L.*) Vetiver (*Vetiveria zizanioides* L.) all on clayey-sandy soil of Plio-Pleistocene age, Altomonto, S. Italy	10.0 2.0 4.0 15.0
Van Beek et al.[2] (2005)	Natural understory vegetation (*Ulex parviflorus* Pourret, *Crataegus monogyna* Jacq., *Brachypodium* var.) on hill slopes, Almudaina, Spain	0.5 – 6.3
Van Beek et al.[2] (2005)	*Vetiveria zizanoides* L., terraced hill slope, Almudaina, Spain	7.5
Mattia et al.[3] (2005)	*Lygeum spartum* L. *Pistacia lentiscus* L. *Atriplex halimus* L. all on eroded badlands in southern Italy	0.3 – 60 3.0 – 20.0 0.2 – 6.0

Norris[2] (2005a)	Mixed grasses on London Clay embankment, M25, England	~10.0
Mickovski et al.[5] (2007b)	*Lolium perenne* L., on agricultural soil	3.0 – 4.5
Deciduous trees		
Endo and Tsuruta[2] (1969)	Silt loam soils under alder (*Alnus* P. Mill.), nursery, Japan	2.0 – 12.0
O'Loughlin and Ziemer[2] (1982)	Beech (*Fagus* sp. L.), forest-soil, New Zealand	6.6
Riestenberg and Sovonick-Dunford[4] (1983)	Bouldery, silty clay colluvium under sugar maple (*Acer saccharum* Marsh) forest, Ohio, USA	5.7
Schmidt et al.[3] (2001)	Industrial deciduous forest, colluvial soil (sandy loam), Oregon	6.8 – 23.2
Danjon et al.[3] (2007)	Mature *Quercus alba* L. on regolithic clays, Georgia, USA	0.01 – 63.0
Conifers		
Swanston[1] (1970)	Mountain till soils under hemlock (*Tsuga mertensiana* Bong. Carr.) and spruce (*Picea sitchensis* (Bong.) Carr.), Alaska, USA	3.4 – 4.4
O'Loughlin[1] (1974)	Mountain till soils under conifers (*Pseudotsuga menziesii* (Mirb.) Franco), British Columbia, Canada	1.0 – 3.0
Ziemer and Swanston[3,5] (1977)	Sitka spruce (*Picea sitchensis* (Bong.) Carr.) - western hemlock (*Tsuga heterophylla* (Raf.) Sarg.), Alaska, USA	3.5 – 6.0
Burroughs and Thomas[4] (1977)	Mountain and hill soils under coastal Douglas-fir and Rocky Mountain Douglas-fir (*Pseudotsuga menziesii* (Mirb.) Franco), West Oregon and Idaho, USA	3.0 – 17.5
Wu et al.[3] (1979)	Mountain till soils under cedar (*Thuja plicata* Donn ex D. Don), hemlock (*Tsuga mertensiana* Bong. Carr.) and spruce (*Picea sitchensis* (Bong.) Carr.), Alaska, USA	5.9
Ziemer[2] (1981)	Lodgepole pine (*Pinus contorta* Dougl. & Loud.), coastal sands, California, USA	3.0 – 21.0
Waldron and Dakessian[4] (1981)	Yellow pine (*Pinus ponderosa*) seedlings grown in small containers of clay loam	5.0

Gray and Megahan[3] (1981)	Sandy loam soils under Yellow pine (*Pinus ponderosa* Douglas. ex Lawson.), Douglas-fir (*Pseudotsuga menziesii*) and Engelmann spruce (*Picea engelmannii* (Parry.) Engelm.), Idaho, USA	~ 10.3
O'Loughlin et al.[2] (1982)	Shallow stony loam till soils under mixed evergreen forests, New Zealand	3.3
Waldron et al.[2] (1983)	Yellow pine (*Pinus ponderosa*) (54 months), laboratory	3.7 – 6.4
Wu[3] (1984b)	Hemlock (*Tsuga* sp.), Sitka spruce (*Picea sitchensis* (Bong.) Carr.) and yellow cedar (*Thuja occidentalis* L.), Alaska, USA	5.6 – 12.6
Abe and Iwamoto[2] (1986)	*Cryptomeria japonica* D. Don (sugi) on loamy sand (Kanto loam), Ibaraki Prefecture, Japan	1.0 – 5.0
Buchanan and Savigny[1] (1990)	Hemlock (*Tsuga* sp.), Douglas fir (*Pseudotsuga*), cedar (*Thuja*), glacial till soils, Washington, USA	2.5 – 3.0
Gray[5] (1995)	*Pinus contorta* Dougl. & Loud. on coastal sand	2.3
Schmidt et al.[3] (2001)	Natural coniferous forest, colluvial soil (sandy loam), Oregon	25.6 – 94.3
Van Beek et al.[2] (2005)	*Pinus halepensis* Mill., hill slopes, Almudaina, Spain	–0.4 – 18.2

1. Back analysis. 2. *In situ* direct shear tests. 3. Root density information and vertical root model equations. 4. Back analysis amd root density information. 5. Laboratory shear tests.

Table 5-3. Lateral root cohesion derived from root area ratio and tensile strength values for different vegetation communities in Oregon, USA (after Schmidt et al. 2001).

Vegetation community	**Lateral root cohesion c'_R (kPa)**
Natural Forest Pit	94.3
Inferred Natural Forest	71.4
Natural Forest Blowdown Landslide	25.6
Industrial Forest Pit	23.2
Natural Forest Landslide	11.0
Industrial Forest Landslide	6.8
Clear-cut Pit	6.7
Clear-cut Landslide	2.7
Herbicided Clear-cut Pit	1.5

The mass of vegetation, surcharge W_v

The mass of vegetation is only likely to have a major influence on slope stability when larger trees (dbh* >0.3 m) are present since the weight of grass, herbs and shrub vegetation is comparatively insignificant. The loading due to a fully stocked forest for tree height between 30 and 60 m, is in the order of 0.5 to 1.5 kPa (Coppin and Richards 1990). A 30 m tall tree having a base trunk diameter of approximately 0.8 m is likely to have a weight of around 100 to 150 kN. Such trees located at the toe of a potential slip could add 10% to the factor of safety (Coppin and Richards 1990). Equally, if located at the top of a potential slip the FOS could be reduced by 10%. Each situation must be individually assessed for the mass of vegetation involved. It should be borne in mind that plant evapotranspiration will reduce the weight of soil as moisture is lost. This effect can be important on slopes of marginal stability.

When larger trees are removed from the toe area of a slope, in addition to the gradual reduction in soil strength due to the loss of evapotranspiration effects, the reduction in applied loading could result in temporary suctions in clay soils which may lead to softening as available water is drawn in to satisfy the suction forces.

Wind loading, D_W

Wind loading is particularly relevant when considering the stability of individual trees but is of lesser significance for general slope stability where the wind forces involved represent a much smaller proportion of the potential disturbing forces and trees within a stand are sheltered to some extent by those at the edge.

Wind forces on single trees may be estimated from Brown and Sheu (1975) and Ancelin et al. (2004) by considering local pressures in relation to wind speed (i.e., $p_s = p\cos^2\beta$ where p_s = wind pressure normal to the tree, p = local wind pressure, β = slope angle). Wind loading on forested slopes may also be calculated by using Equation [6]:

$$p = 0.5\rho_a V^2 C_D \tag{6}$$

where p = wind pressure, ρ_a = air density in kg/m^3, V = wind velocity in m/s and C_D = dimensionless drag coefficient (Hsi and Nath 1970). Average wind speeds for Europe may be assumed from the wind resources map (Troen and Petersen 1989).

Soil strength increase due to moisture removal by roots, c'_s

Observations of moisture deficit around trees due to the effects of evapotranspiration and the problems this has caused for buildings and

---

* - diameter at breast height

Suctions and changes in pore water pressure due to vegetation, u_v

The moisture content and pore water pressures within a slope are closely related. Suctions or changes in pore water pressure can be measured over the long-term through the installation of tensiometers. Tensiometers installed on slopes are able to monitor and record the response of the ground suctions to rainfall events and periods of wet or dry weather (Greenwood et al. 2001). Indraratna et al. (2006) carried out numerical modelling of the matric suctions of native Australian vegetation used for stabilising railway corridors built over expansive clays and compressive soft soils. Indraratna et al. (2006) showed that the vegetation improves the shear strength of the soil by increasing the matric suction, and as a result curtailing slope movements.

Tensile root strength contribution, *T*

The tensile strengths of roots of various diameters from different species have been measured in the laboratory and found to be typically in the order of 10 – 40 MPa (see Chapter 4).

In the field, to make use of the available tensile strength to enhance slope stability the root must have sufficient embedment and adhesion with the soil. The available force contribution from the roots can be measured by *in situ* pull out tests using hand digital force gauges or mechanical/hydraulic jacking apparatus (Figure 5.7, see Norris and Greenwood 2000, 2003 for procedure).

Figure 5-7. Root pull out apparatus (Photo: J.E. Norris).

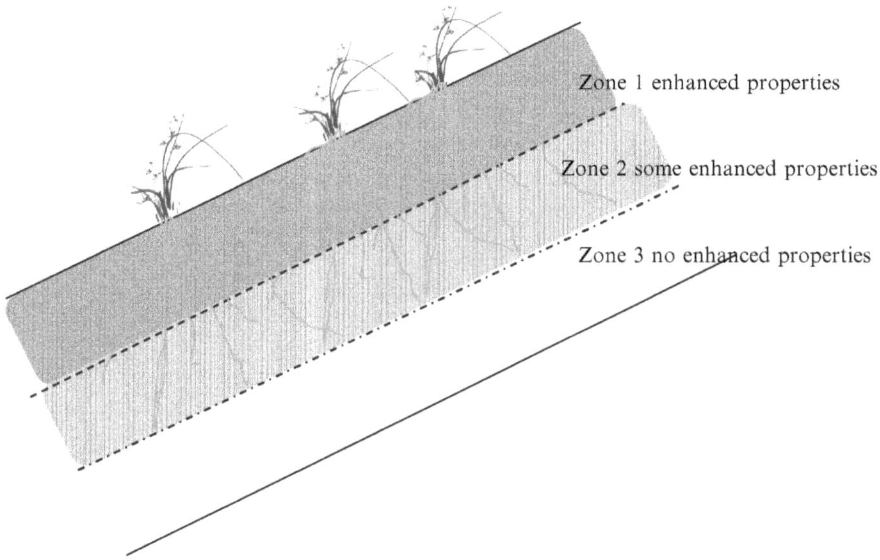

Figure 5-8. Zones of enhanced soil properties for grass and shrub vegetation cover (modified from Greenwood et al. 2003).

The maximum breaking force or pull out resistance of the roots and the associated root area ratio (root size and distribution) is used to determine the appropriate root reinforcement values for inclusion in Greenwood's General equation. The distribution of roots in a vertical trench wall profile of soil can be assessed by measuring the Root Area Ratio (RAR), i.e., the proportion of the cross-sectional area (CSA) of a sample section of soil that is occupied by roots.

The available root force acting on the base of the slice of the analysis, T, can be estimated by introducing the term T_{rd}, the available (design) root force per square metre across a particular plane (for example, the slip surface) within the soil. Values of T_{rd} may be assigned for different root zones evident beneath the ground surface (Figure 5.8). T_{rd} is based on the ultimate root force available across the plane considered, T_{ru} in kN (per square metre of soil), with a suitable safety factor due to the roots, FOS_r applied, i.e.,

$$T_{rd} = \frac{T_{ru}}{FOS_r} \tag{7}$$

T_{ru} may be estimated based on the observed or assumed root distribution and determination of characteristic resisting forces for the roots of varying diameters by root pull out and tensile strength testing (Norris and Greenwood 2000, 2003; Greenwood et al. 2004; Norris 2005a).

The natural evolution of plant roots is such that they are generally just sufficient to serve their purpose of maintaining stability against gravitational and wind forces. It has been observed that the pull out resistance of a root is likely to be only slightly less than the measured tensile strength of the root (Norris 2005b). The tensile strength of the root is therefore likely to be a reasonable indicator of the maximum pull out resistance available.

There is considerable uncertainty about root distribution in the ground and the resisting forces which are available in particular soil conditions. For this reason a high estimated value of FOS_r is recommended. Values of FOS_r of 8 or 10 are currently used to reflect the uncertainties and to allow for the large strains, typically in the order of 20%, necessary to generate the ultimate root resistance to pull out (Greenwood et al. 2004). It may be possible to reduce the FOS_r as the root zones around the plant or tree are better characterised on a seasonal basis and more root pull out information becomes available.

T_{rd} may therefore be estimated based on the measured pull out strengths or as a proportion of the measured or assumed tensile strength of the roots crossing the slip plane.

$$T_{rd} = \frac{assigned\,ultimate\,root\,resistance \text{ x } root\,area\,(per\,sq.m.of\,soil)}{FOS_r} \tag{8}$$

The force T applicable to a slice of the stability analysis is given by Equation [9].

$$T = T_{rd} l \tag{9}$$

where l = the length of slip surface affected by the roots (assuming unit width of slope).

1.2.1 Stability analysis to include the influences of vegetation

The influences of vegetation on the FOS of a slope can be modelled by routine limit equilibrium stability analysis methods, e.g., the method of slices. Two methods of analysis (Greenwood's and Fellenius') are readily adapted for including the influences of vegetation. The addition of these influences of vegetation in Bishop, Janbu and other more sophisticated published solutions where the global FOS is applied to the shear strength parameters for each slice of the analysis results in unrealistic force scenarios for the slices where anchor and reinforcement loads are applied (Krahn 2001).

The Greenwood General equation (Greenwood 1989, 1990, 2006; Morrison and Greenwood 1989) is considered particularly appropriate for including vegetation because it takes full account of hydrological (seepage) forces to give a realistic estimate of the FOS for all types of slopes and slip surfaces:

$$FOS = \frac{\Sigma\left[c'l + (W\cos\alpha - ul - (U_2 - U_1)\sin\alpha)\tan\phi'\right]}{\Sigma W \sin\alpha} \tag{10}$$

where c' = effective cohesion at base of slice, l = length along base of slice, W = weight of soil, α = inclination of base of slice to horizontal, ϕ' = effective angle of friction at base of slice, u = water pressure on base of slice, U_1 and U_2 = interslice water forces on left and right hand side of slice.

The interslice water forces, U_1 and U_2, may be calculated based on assumed hydrostatic conditions below the phreatic surface or derived from a flow net for more complex hydraulic situations. It should be noted that if the interslice forces U_1 and U_2 are equal the equation becomes:

$$FOS = \frac{\Sigma\left[c'l + (W\cos\alpha - ul)\tan\phi'\right]}{\Sigma W \sin\alpha} \tag{11}$$

Equation [11] is the well known Fellenius equation (see Table 5-1) which is appropriate to use for a planar, slab slide on a continuous slope with seepage parallel to the slope. However the user should be cautious as in practice, the parallel seepage is often interrupted by less permeable layers resulting in a local reduction in the FOS. The actual hydraulic conditions are therefore more correctly modelled using the Greenwood General equation (Morrison and Greenwood 1989).

The simple mathematical form of the Greenwood equations with the FOS simply expressed by a summation of restoring and disturbing moments or forces makes the inclusion of additional forces due to ground reinforcement, anchors or plant roots relatively straightforward (Equation [12]):

$$FOS = \frac{\Sigma\left[(c' + c'_R)l + ((W + W_v)\cos\alpha - (u + \Delta u_v)l - ((U_2 + \Delta U_{2v}) - (U_1 + \Delta U_{1v}))\sin\alpha - D_W\sin(\alpha-\beta) + T\sin\theta)\tan\phi'\right]}{\Sigma\left[(W + W_v)\sin\alpha + D_W\cos(\alpha-\beta) - T\cos\theta\right]} \tag{12}$$

It is noted that the tangential reinforcement force, $T\cos\theta$, in Equation [12], is correctly deducted from the denominator as it is a negative disturbing force. In practice the term is often assumed to be a positive restoring force and is added to the numerator. This approach is statically correct in accordance with the force diagram. The differences in the calculated FOS by either approach are small with identical values calculated when FOS = 1.

Whilst the FOS in Equation [12] is expressed as a traditional ratio of restoring to disturbing forces, the equation may be adapted to include partial factors on each individual term in accordance with European codes of practice, Eurocode 7 (BS EN 1997-1 2004; BS EN 1997-2 2007).

Computer packages

A Microsoft Excel spreadsheet, known as 'SLIP4EX' (Greenwood 2006), was developed to compare routine methods of analysis for a given slip surface and to quantify the changes to the FOS due to the influences of the vegetation. This program is available from the author john.greenwood @ntu.ac.uk. Other computer software packages are available for slope stability analysis, e.g. Slope-W (http://www.geo-slope.com/), and STABL (http://www.ecn.purdue.edu/STABL/).

The energy approach

The energy approach was developed by Ekanayake et al. (1997) and Ekanayake and Phillips (1999a,b, 2002), to take into account the contribution of roots to soil strength for specific New Zealand soils. The method allows for the fact that roots can withstand large-strains during displacement of the soil-root system. To enable this method to be applicable to all cases, the original energy approach is generalised and a soil-water infiltration model is introduced.

In the stability analysis, the method incorporates the ability of tree roots to withstand strain during shear displacement. The characteristics of the shear stress–shear displacement curve obtained from an *in situ* direct shear test are used to find the total energy capacity of the soil-root system and the amount of energy exchanged up to the current displacement (Figure 5.9). The energy exchanged during the shearing process is directly related to the area between the stress-displacement curve and the *x*-axis. The total energy capacity of the soil-root system is the area under the soil with roots curve up to the shear displacement at peak shear stress.

The energy approach stability analyses method estimates the FOS using the energy associated with the root-soil shearing process. The FOS is defined by the ratio of energy already spent, up to the current shear displacement and the total energy capacity of the soil-root system. As the shear displacement is taken into account within the energy approach, this method will always overestimate the FOS compared to limit equilibrium methods.

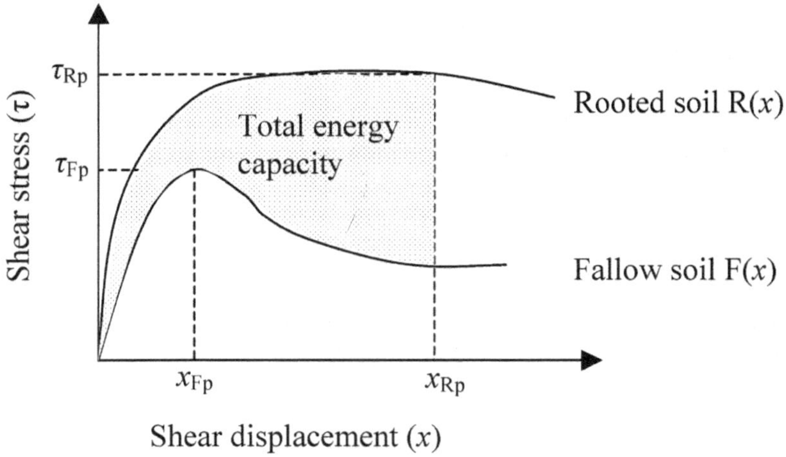

Figure 5-9. Ideal shear stress–displacement curves for fallow soil F(x) and soil with roots R(x). x_{Fp} is shear displacement at the peak stress (τ_{Fp}) for fallow soil and x_{Rp} is shear displacement at the peak stress (τ_{Rp}) for soil with roots. The shaded area between the two curves represents the total energy capacity of the soil-root system (after Ekanayake and Phillips 1999b).

Finite element models

Finite element modelling is based on a numerical approximation solution for solving problems represented by partial differential equations. The 'problem' or model is divided into discrete elements, each element is connected by nodes at the corners which form triangular or quadrilateral shapes. The behaviour of unknown variables is modelled at the nodes through appropriate polynomial equations. Two finite element packages which can be used to model vegetation and soil behaviour are PLAXIS and FLAC.

PLAXIS is a finite element package specifically intended for the two dimensional analysis of deformation and stability in geotechnical engineering projects (Brinkgreve 2002). Geotechnical applications require advanced constitutive models for the simulation of the non-linear, time-dependent and anisotropic behaviour of soils and/or rock. In addition, since soil is a multi-phase material, special procedures are required to deal with hydrostatic and non-hydrostatic pore pressures in the soil. PLAXIS can model the complex interaction between geotechnical structures and the soil.

The program allows for graphical input of geometry models, automatic mesh generation and 15-node triangular elements to model the deformations and stresses in the soil. Soil behaviour can be modelled using the Mohr-Coulomb model, advanced soil models such as the 'soil hardening' model, or other user-defined soil models (see Fredlund and Rahardjo 1993). Vegetation can either be modelled as geogrids for grass root networks, or as a series of anchors to replicate tree roots.

FLAC is a commercially available finite difference code with widespread application in geo-engineering (Itasca 2002). It mimics the stress-strain behaviour numerically so the strain-dependent effect of reinforcement can be simulated more realistically with fewer simplifying assumptions. Moreover, the root reinforcement model in FLAC offers the user to specify varying root and soil properties along the slope and the influence of the hydrology on the effective stress can be evaluated rigorously. This is highly advantageous since root reinforcement is influenced by the type and nature of the vegetation and local variations in soil conditions. An example of the use of FLAC2D to model root reinforcement can be found in van Beek et al. (2005).

2. HAZARD ASSESSMENT OF SOIL EROSION

2.1 Introduction

Soil erosion by water and wind affects both agriculture and the natural environment, and is one of the most important (yet probably the least well-known) of today's environmental problems (http://soilerosion.net/).

Soil erosion is an important issue and it concerns large areas of the terrestrial environment. It has a large economic impact as it degrades the most fertile part of the soil which negatively affects crop productivity (on-site effect) on the eroded areas and creates off-site problems, e.g., silting up of reservoirs. We should distinguish wind erosion from water erosion, as both processes are quite different both in process and their area of occurrence.

The occurrence of erosion is related to:

- rainfall characteristics (erosivity)
- soil material (erodibility)
- vegetation cover
- relief

Rainfall is more effective as an erosive factor when its intensity is high. High intensity rainfall events are mainly found in the Mediterranean, sub-tropical and tropical climate zones whereas in temperate zones these events are far less frequent.

In semi-arid and arid environments erosion is dominated by wind activity. Figure 5.10 shows the rainfall regimes under which both erosion types are dominant.

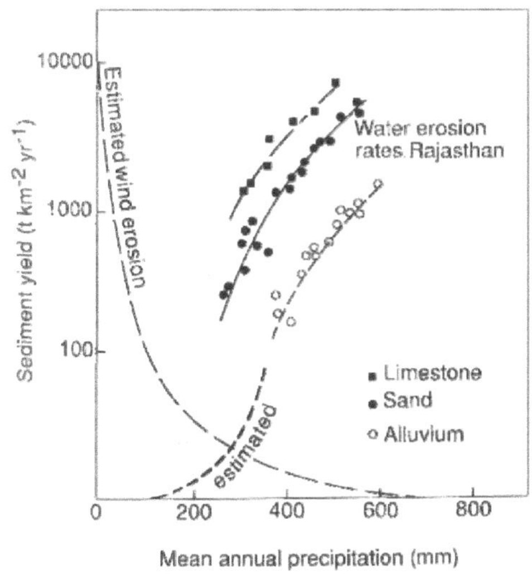

Figure 5-10. Measured and estimated rates of erosion by wind and water in different climatic conditions. From Cooke et al. (1993), reprinted by permission of the publisher.

Soil material

Porous and permeable materials are less susceptible to water erosion than finer textured soils. Silt and clayey soil may show high erodibility, although this latter factor is also influenced by soil organic carbon levels and soil mineralogy. Sandy soils may however be very vulnerable to wind erosion when organic matter is almost absent, or when water repellence is important.

Vegetation cover acts as a protective factor for the soil. It reduces the kinetic energy of the falling rain drops on the soil and it also promotes infiltration of water in the soil. Furthermore it also reduces overland flow velocities enhancing infiltration. Arable lands devoid of vegetation after ploughing can be extremely vulnerable to erosion.

Relief and terrain characteristics determine the slope gradients, slope curvature and slope length which all influence soil erosion. Steep slopes are more vulnerable to water erosion as well as long slopes. Areas with a long wind fetch are more vulnerable to wind erosion.

A broad discussion on these topics can be found in excellent textbooks on soil erosion such as that of Hudson (1979) and Morgan (2005).

2.1.1 Techniques of soil erosion assessment

Erosion can be assessed in many ways and a range of methodologies have been developed. These range from simple surveying techniques, long-term erosion measurement experiments, short intensive simulation experiments or GIS and remote sensing analysis. Assessment depends on the goal, and the time and money available as to which methodology can be applied. An excellent overview of erosion assessment and measurement is the work of Hudson (1993). This document is recommended by the authors as only a brief description is given of the main groups of methodologies that can be applied in the following text.

A general difference should be made between surveying techniques, which are more descriptive, but can be applied to larger areas and measuring techniques, which are more suitable to assess actual rates of erosion. In the first case, a good knowledge of the landscape and soils is necessary whereas in the last case, one should be fully aware that fine scale measurements cannot directly be extrapolated to larger areas as each process acting on the landscape has its own spatial and temporal process-domain, thresholds are involved in the geomorphic and hydrological response and connectivity between landscape units rules the movement of soil material through the landscape (Cammeraat 2004).

The use of erosion models is tempting but to be able to work with calibrated erosion models measured field data are necessary. Simple erosion models such as the empirical Universal Soil Loss Equation (USLE) (Wischmeier and Smith 1978) are often used, but have their limitations as they are developed or calibrated for specific conditions, e.g., for the USLE: slopes $< 6°$; agricultural land and calibrated in standard bounded plots.

2.1.2 Surveying methods

Soil profile truncation

Soil erosion can be assessed from studying the development of the soil profile. The soil profile normally has a set of horizons that develop over long periods of time. When soil formation rates and or weathering rates are equal or larger than the soil erosion rate, soil profiles remain *in situ*. In the reverse case, soils will lose their upper soil horizons. Soils lacking a B and/or an A horizon are clear field indicators of accelerated erosion rates, which is often related to agricultural activity on sloping areas.

A survey of truncated soils may give a good indication of the spatial distribution of eroded soils and might help in determining the most affected areas or pinpointing areas at risk. A good knowledge of field pedology is prerequisite for applying this method.

Colluviation

Soils removed from sloping areas by soil erosion processes are often deposited at the foot of the slope in thick layers. Colluvial deposits can be recognized by the increased presence of organic matter, sometimes with an organic matter enriched layer of soil, often associated with charcoal fragments and a dirty coating around soil particles. Furthermore soil profile development is retarded because of the high deposition rate of colluvial material. As colluviation is often associated with soil profile truncation, field knowledge of soils is indispensable.

Soil surface properties

Careful observation of the soil surface is a good methodology to assess the occurrence of soil erosion processes. In Australia some interesting manuals have been published which enable the assessment of erosion and degradation of rangelands and grass areas under semi-arid conditions (Tongway 1994; Tongway and Hindley 1995). These methods can be good starting points to apply similar methodologies in other environments in combination with, for instance, indicator techniques (Imeson and Cammeraat 1999).

Surface wash can be observed by several indicators, for example, the exposure of lateral tree roots (Figure 5.11), and the presence of trees or shrubs standing on small mounds.

Slaking and Crusting is another important feature indicating reduced infiltration rates and erosion sensitive soils. Many different types of crusts exist which are well described in Casenave and Valentin (1989) for semi-arid environments or in Valentin and Bresson (1992) for soils in temperate climates.

Rilling when present is a clear sign of flow concentration with high soil material transport capacities. This type of erosion can easily by aggravated and lead to the formation of large gullies (Figures 5.12 and 5.13).

Tillage erosion is the result of tillage of soils on sloping areas, which causes a net downward transport of soil material (Quine et al. 1999; Takken et al. 2001). In upper slopes this can be seen from trees standing on isolated small hills and in lower slopes, trees might by partially covered at their base.

Figure 5-11. Sheet wash erosion in the Lake Baringo District in Kenya (Photo: E. Cammeraat).

Figure 5-12. Rill and gully erosion in the Lake Baringo District in Kenya (Photo: E. Cammeraat).

Figure 5-13. Rill erosion induced by ploughing (Guadalentin basin, Spain) (Photo: E. Cammeraat).

2.1.3 Measuring methods

Changes in soil surface levels

Changes in soil surface levels can be estimated by the use of erosion pins. Small pins are inserted in the ground, in such a way that they are permanently fixed and not subjected to vertical or lateral movement (soil shrinkage, creep). By measuring the height difference between the soil and the soil surface, soil surface lowering can be followed. Errors can be obtained by the influence of the pins themselves as they block air and water flow and the hydraulic regime around the pin is different compared to the open surface. Haigh (1977) discusses the possible errors resulting from applying this method.

In semi-arid environments, trees or shrubs may be seen standing on isolated small hills which could also be a sign of soil erosion, as the vegetation protects the surroundings from splash erosion. In other cases, this might indicate concentrated flow around vegetation clumps where plant roots protect the soil from water erosion.

A more modern method to determine the spatial distribution of erosion is the determination of the spatial pattern in the presence of radioactive nuclides like Caesium-137 derived from radioactive fall out (Walling and Quine 1990; Morgan 2005).

Measuring rill or gully erosion

The presence of rills and gullies in the landscape reflect also the activity of soil erosion processes. This activity can be estimated by the presence of or lack of vegetation, soil crusting and cryptogamic crusts. When well-established vegetation is present in a rill or gully (head) wall this indicates that it is not very active. Also, the presence of cryptogamic crusts indicates rather stable surfaces.

The development of gullies or rills may be followed over time. Measurements can be performed by placing a grid of reference markers in the surroundings of the gully (Hudson 1993). Measuring the distance between the gully head or wall to the reference points can give an indication of their growth. An indication of volumetric change and extension can be determined when the depth of the gulley is monitored.

Rills and gullies often occur in agricultural soils but are in most cases ploughed away by the farmer. In these areas, rilling and gullying is often associated with the direction of tillage (Figure 5.13). Erosion may increase enormously when contour ploughing is not applied.

Actual rates of erosion can also be determined by measuring the sediment output of a rill or gully in the same way as described below.

Measuring surface erosion

Erosion plots can be built to measure erosion rates. A soil surface is selected and the runoff and sediment produced by the area is collected in a gutter or trough. The plot can be bounded which is normally performed using the argument that the rate can be coupled to a fixed surface. However in reality this is usually not the case as the runoff and sediment are often not originating equally from the whole plot, but normally originates more from the area near the gutter. Long term experiments might suffer from sediment depletion as well. Bounding of the plot also limits the slope length, which is an important factor and it also excludes water coming from higher upslope to reach the gutter. However, many experiments use standardized bounded plot dimensions after the highly influential field experiments carried out in the US to support the USLE (Wischmeier and Smith 1978). Open (non-bounded) plots are also used and are more adjusted to the natural catchment areas present within a slope, but this deserves a more detailed topographic survey of the actual watershed that is drained by the gutter or troughs. In this case the origin of the water is also not clear due to the strong heterogeneity of soil surfaces.

Sediment can be sampled continuously during events by hand or with instruments, e.g., automatic samplers or turbidity meters, or on an event base.

Retention basins or catchpits. When small basins are present downstream of an eroding area, the amount of sediment delivered by this area can also be estimated from the soil trapped in small retention basins (Verstraeten and Poesen 2000). These are currently increasingly built to remediate off-site effects of erosion in sensitive areas but can also be designed especially for assessment purposes.

Rainfall simulations are often applied to measure erosion or runoff from soil surface areas. Rain in semi-arid environments does not occur frequently and intensity and amounts are unpredictable and variable. These problems can be overcome by rainfall simulation experiments (Figure 5.14). They have the advantage that they can be carried out under controlled conditions with regards to rainfall intensity and duration. Normally, rainfall is simulated over a plot where runoff and sediment are collected in a gutter or a trough. The big disadvantage of rainfall simulators is however, that the terminal velocity of the raindrops falling on the surface is critical with regards to their kinetic impact on the soil surface. Mostly, rainfall simulators are much lower than 9-10 m, which is normally the height for a drop to attain its terminal falling velocity. In particular, dripping plate simulators have this problem, e.g., Bowyer-Bower and Burt (1989). Simulators with nozzles have higher drop velocities as these drops are being produced under higher pressures. The spatial heterogeneity of the rainfall depth of simulators may also cause a problem (Lascelles et al. 2000). Upscaling is in any case a problem when working with fine scale measurements, as the erosion response is highly non-linear and complex, with different processes being dominant at different scales.

Figure 5-14. The drip-plate rainfall simulator (Amsterdam-type. Photo: E. Cammeraat).

Remote sensing and computer simulation methods

Many methods exist to predict erosion from fields or catchments using simulation models. As this topic is outside the purpose of this book, it is only briefly described and only one method is referred to from the vast literature on this topic. The most well known model is the USLE model which is simple and has been successfully applied on many agricultural soils (Wischmeier and Smith 1978). However it is not suitable for erosion assessment for larger areas such as watersheds (Wischmeier 1978). Many other soil erosion models exist on many different scales but they all highly depend on input data, which are often difficult to obtain.

Remote sensing is also increasingly used, by the interpretation of surface topography changes from aerial photography or by geodetic processing of high quality aerial photographs, e.g., Vandaele et al. (1996).

Change in topsoil properties can also be detected from spectral properties of soil surfaces and this can also be applied in regions where bare areas are present with characteristic differences in reflectance and spectral properties between the different soil horizons exposed, e.g., Metternicht and Fermont

(1998), Hill and Schütt (2000). Combining the results from both remote sensing and GIS is increasingly carried out.

3. STABILITY OF VEGETATION ON SLOPES

The stability of vegetation on slopes, especially forested slopes, is equally as important as the stability of the soil that the vegetation is planted in. This section reviews the hazards of wind and snow damage on forested slopes.

3.1.1 Windthrow Hazard

The practical problems and economic costs that result from windthrow of trees (Figure 5.15) has stimulated much research into tree root anchorage. This research effort is almost inseparable from the related topic of stabilisation of soil on slopes by tree roots. Much research on anchorage has focussed on the nature of the root-soil bond (for example, Waldron and Dakessian 1982; Operstein and Frydman 2000; Mickovski et al. 2007a). However, the effects of trees on soil stability are more complex than this. Trees provide considerable protection to slopes by sheltering the slope surface from the direct effects of wind and rain, by extracting soil water through transpiration, and by holding soil on both fine and coarse roots (Keim and Skaugset 2003). To maintain these benefits in forested slopes that are actively managed, consideration should be given to minimising windthrow at all stages during planning, managing and harvesting.

3.1.2 Soil loss from windthrow on slopes

Tree uprooting on slopes can lead to pits forming in the soil, in which water collects and infiltration is increased. However infiltration is not the only process leading to soil loss following windthrow. An investigation by Nicoll et al. (2005) predicted that for dense forest stands on steep slopes, where windthrow overturns root plates downslope, the potential downslope displacement of soil is in the order of 1800 m^3ha^{-1} from the displaced soil-root plates alone, even before additional soil is displaced by erosion processes associated with pits. This rate of soil loss is more than 1000 times the rate expected from standard forestry operations. As soil loss must be considered as an almost permanent degradation of the site, with considerably

Figure 5-15. Windthrow of plantation trees on a hill side in Scotland. Photograph courtesy of the Forestry Commission, UK.

greater long-term consequences in terms of forest sustainability than windthrow, soil conservation should become the primary consideration on such sites.

Nicoll et al. (2006) showed that species choice, soil type and rooting depth all influence anchorage. Therefore, these criteria may be used in any risk analysis to decide how forest stands should be designed, established and managed on steep slopes. Species with relatively good predicted anchorage or slow growth may be chosen for such sites, and the suitability of silvicultural treatments to be applied to them should be assessed based on the risks of windthrow and resulting soil loss. For example, particular care should be taken in applying thinning treatments or in respacing on vulnerable slopes (see Chapter 7).

3.1.3 Assessment of windthrow hazard

There are three basic approaches to the assessment of windthrow hazard: observational, mechanical and empirical (Cucchi et al. 2005; Mickovski et al. 2005). These are used either independently or in combination with each other:

- **Observational** approaches use a checklist of indicators.
- **Mechanical** approaches predict the critical wind speed for over-turning from winching and wind tunnel studies, and the probability of critical wind speed from wind mapping/modelling work.

- **Empirical** approaches use regression techniques to predict the probability of damage as a function of environmental and management variables.
- **Combined** approaches incorporate elements of the observational, mechanical and empirical approaches.

The wind risk system 'ForestGALES' (Geographical Analysis of the Losses and Effects of Storms in Forestry) is an advanced example of the combined approach. It was developed for conifer plantations, and is based on winching tests, wind tunnel studies, information on tree and soil characteristics, site wind exposure and wind climate (Quine and Gardiner 1998; Gardiner et al. 2004). The output gives the probability of damage to a stand over time. ForestGALES was designed for UK forests but has been adapted to work in parts of France, Denmark, Canada, Japan and New Zealand. It is adaptable for other countries, depending on availability of data on tree anchorage and wind climate. The ForestGALES decision support system is used by managers to minimise windthrow risk whilst optimising economic returns from timber. To do this, the manager must decide what level of risk he or she can accept and must always be prepared to accept some loss through windthrow.

Another method, which has been used in British Columbia, Canada, is based on the observational approach, but includes some elements of the empirical approach. This system uses windthrow risk assessment field cards to evaluate the windthrow risk (Mitchell 1998). In general, windthrow risk for an individual tree is a function of biophysical risk caused by the environmental factors and the treatment risk arising from the management factors. The environmental factors affecting windthrow are broadly grouped into topographic exposure, soil and stand properties, whilst management factors include the silvicultural management strategies (treatments) that cause change in wind loading on residual trees after the treatment.

ForestGALES and the British Columbia system are further described in Section 3.1.4.

Topographic exposure

Topography influences wind flow and, in turn, the vulnerability of trees to windthrow (Table 5-4). It takes into account the position of a single tree or a stand relative to prevailing winds. After the initial deceleration close to the ground upwind of ridges or hills, winds accelerate over their crests and often create separation bubbles behind them (Figure 5.16).

structures are well documented (e.g., Hunt et al. 1991; Biddle 1998). However when it comes to relying on tree and shrub roots to remove water and hence strengthen soil slopes it is not quite so straightforward. Vegetation trials on the M20 motorway, U.K., indicated large seasonal variations in moisture content (and hence the undrained soil strength) of the south facing trial area. These seasonal variations masked any effects the vegetation may have contributed to increased soil strength (Greenwood et al. 2001).

During particularly wet periods, the ability of plant roots to influence the seasonal moisture content will be curtailed and therefore any enhanced soil strength gained previously by evapotranspiration will be reduced or lost entirely to an extent difficult to quantify. Hence this effect cannot be taken into account at such critical times. However, it can be assumed that there is a narrowing of the window of risk of failure due to soil saturation by storm events or periods of prolonged rainfall. Furthermore, whilst moisture content changes influence the undrained shear strength (c_u) the effective stress parameters (c' and ϕ'), as generally used in routine stability analysis, are not directly influenced by the changing moisture content, although the water pressures (suctions) used in the analysis may well be.

It should be borne in mind that desiccation cracks, possibly extended during dry periods by the presence of certain vegetation, will encourage a deeper penetration of water and water pressures into the soil during wet periods. However, these cracks will subsequently provide pathways for roots to extend deeper into the soil in their search for moisture and nutrients. Vegetation may also promote unwanted desiccation cracks on highway roads (Figure 5.6).

Figure 5-6. Embankment shrinkage due to the presence of high water demand trees (mainly oaks) on the overbridge at Junction 12, M11, U.K. (Photo: Courtesy of C. Bull, URS Corporation Ltd, Bedford, U.K.).

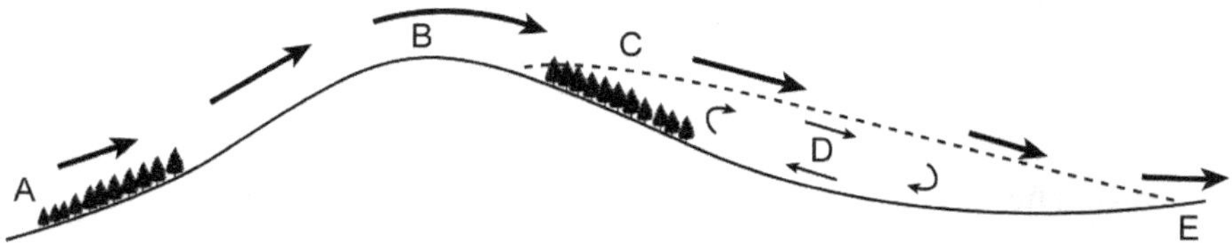

Figure 5-16. Features of the airflow over forested hills. A: presence of forest on lower slopes reduces wind speed at top; B: speed-up of the wind at summit; C: separation of flow in lee of hill encouraged by presence of trees; D: slack air in lee of hill; E: reattachment of flow downstream of hill (after Quine et al. 1995).

Table 5-4. The effect of tree/stand position and the prevailing wind direction on the vulnerability (low, moderate or high) to windthrow (adapted from Alexander 1987).

Topographic position of the tree or stand	Wind direction	
	Parallel	**Perpendicular**
Flat	Moderate	Moderate
Slope toe	Moderate	Moderate
Slope crest	High	Moderate
Knoll	High	Moderate
Side slope	Moderate	Moderate
Ridge	High	High
Shoulder	High	High
Saddle	High	High
Sheltered valley	Low	High

Simple assessments of topographic exposure can be made using Topex (Miller et al. 1987), which implies that the windiness of a site can be assessed with regards to its environment. For example, a slope aspect perpendicular to the prevailing wind direction is particularly exposed, but a valley parallel to prevailing winds may experience even higher wind speeds due to the funnelling effect.

Topex is calculated by summing the angle to the sky line at the eight principal cardinal points. High values indicate the presence of higher ground near the measurement site, and therefore the site is considered to be sheltered. These values are incorporated into the DAMS (Detailed Aspect Method of Scoring) system used in the UK as a measure of site windiness (Quine and White 1993). DAMS combines scores depending on region of the country (i.e., the wind zone of the location), elevation, Topex, aspect and funnelling.

Stand properties

Tree height

It has long been recognised that windthrow risk tends to increase with an increase in tree height (Cremer et al. 1982; Savill 1983; Miller 1985). Cremer et al. (1982) links this to three factors:

- An increase in stem height implies an increase in the turning moment applied to the base of the stem.
- Because wind speed increases with height inside and above the canopy, trees that are taller than their neighbours are more vulnerable.
- Trees in fully stocked stands have a decreasing diameter to height ratio as they grow, meaning that they are less tapered and hence more vulnerable to breakage or uprooting.

Irregular stand structure

Several empirical studies have investigated the effect of irregular stand structure on the risk of windthrow (Lanier 1994; Schütz 1997; Otto 2000; Dvorak et al. 2001). Mason (2002) reviewed these reports and found that although irregular stands are widely believed to be less vulnerable to wind damage, the many confounding factors, including site and topographical variation mean that this assumption may not always be correct. The ForestGALES model was used to assess windthrow risk in simulated irregular Sitka spruce stand conditions (Mason 2002). The main conclusion from this work was that the lower height over diameter (H:D) ratios of dominant trees, which is a widely recognised characteristic of irregular stands, helps improve tree and stand stability against wind damage. However, the extent of the increase in stability is mediated by site characteristics and by local wind climate. An effect perhaps more important than an increase in windthrow resistance is the greater plasticity of irregular stands. The faster recovery of wind-damaged irregular stands to their desired state was shown by Brang (2001) for protection forests in the Alps. This is why the risk of 'extensive' wind damage is considered to be lower in irregular, or uneven-aged, stands.

Existing damage in a stand

Signs of existing damage within stands can be indicative of the stand reaching a critical stage. Apart from obvious signs of blown or snapped trees, this can be indicated by evidence of pumping around trees (areas of wet ground-up soil on the surface where the tree is rocking), signs of extensive decay (rotten stems, fungi on stem), and compression creases in the bark of the tree.

However, if the damage is clearly associated with a specific localised problem, such as flooding caused by a spring or blocked drain or damaged roots or stems following harvesting operations, the stand may not be as vulnerable as the damage suggests. Evidence from studies in commercial plantations suggests that small windthrow gaps can remain with little expansion for many years under many circumstances (Quine 2002).

Windthrow at margins

An untreated forest edge is an abrupt barrier presented to the wind, and the edge trees are subjected to severe wind loading. The edge disrupts the flow for a distance of approximately 4-5 tree heights downwind at which point the flow direction is into the top of the forest and the trees are more vulnerable (Gardiner et al. 2005; Yang et al. 2006). This is where the gustiness of the wind suddenly increases, and where tree-scale damaging gusts have fully developed. If the edge trees are removed from a stand, for example, when widening a road, the remaining stand without the protection of large, windfirm edge trees, becomes particularly vulnerable to windthrow and damage is commonly observed even with relatively low wind speeds.

Windthrow and spacing

Similarly, the risk of windthrow increases after thinning as wind load on individual trees is increased and their capacity to dissipate energy by crown contact is decreased (Cremer et al. 1982; Savill 1983). It is considered that the effect is maximal immediately after the operation and then decreases with time (Lohmander and Helles 1987), as the trees adapt their growth in response to the wind, called "acclimative growth" (see Chapter 4) and thereby strengthen their anchorage (Nicoll and Ray 1996). Depending on its vigour, the stand may recover as soon as 2 – 5 years (Cremer et al. 1982; Savill 1983) but recovery times as long as 15 years have also been reported (Busby 1965).

The effect of initial spacing or early thinning is not as clear. Many authors consider that, through an increase in stem taper (or H:D ratio), wide spacing increases the stability of a stand (Cremer et al. 1982; De Champs 1987; Blackburn and Petty 1988; Galinski 1989; Maccurrach 1991; Valinger et al. 1993; Peltola and Kellomaki 1993). However, this conclusion was put into perspective by Gardiner et al. (1997) who showed that the evidence for an increase in stability was reasonable in relation to stem breakage but weak in relation to overturning. Gardiner et al. (1997) showed that with increased spacing, the bending moments transferred to the base of the stems increased faster than their capacity to resist them.

3.1.4 Windthrow Hazard Models

ForestGALES Model Description and Development

ForestGALES is a mechanistic model designed to replace the Windthrow Hazard Classification formerly used by the forest industry in the UK (Miller 1985; Gardiner and Quine 2000; Gardiner et al. 2004). The program calculates the critical wind speed to cause damage to a stand and the return period for that damage to occur. The use of such a model creates more flexibility for testing different forest management scenarios such as choice of cultivation, thinning options, drainage improvements, the impact of clearfellings, or the creation of retentions.

ForestGALES calculates the wind forces on trees within forest stands as a function of the tree characteristics. Firstly the model calculates the threshold wind speeds required for overturning and breakage as a function of tree height, diameter, current spacing, soil type, cultivation, drainage and choice of species (Gardiner et al. 2000, 2004). The average wind loading on each tree is calculated from the stress imposed on the canopy by the wind from a calculation of the aerodynamic roughness (z_0) and the zero plane displacement (d).

The resistance to breakage is based on the calculation of the bending moment required to cause the stress in the outer fibres of the stem to exceed the Modulus of Rupture (MOR) of the wood. It is possible to write an equation [13] to give the critical wind speed at canopy top for breakage:

$$uh_{break} = \frac{1}{kD}\left[\frac{\pi MOR \times dbh^3}{32\rho G(d-1.3)}\right]^{\frac{1}{2}}\left[\frac{f_{knot}}{f_{edge} f_{CW}}\right]^{\frac{1}{2}} \ln\left(\frac{h-d}{z_0}\right) \tag{13}$$

where $k = 0.4$ is Von Karman's constant, D is the average spacing between trees, G is an empirically derived gust factor, dbh is diameter at breast height, ρ is density, and h is mean tree height. The factors f_{knot}, f_{edge}, and f_{CW} account for the reduction in wood strength due to knots, the position of the tree relative to the edge and the additional load due to the overhanging weight of the crown respectively.

The resistance to overturning has been obtained from tree pulling experiments on almost 2000 trees (Nicoll et al. 2006) and is found to be strongly related to stem weight. A similar equation to Equation [13] can be derived for the critical wind speed at canopy top for overturning:

$$uh_{over} = \frac{1}{kD}\left[\frac{C_{reg}SW}{\rho Gd}\right]^{\frac{1}{2}}\left[\frac{1}{f_{edge}f_{CW}}\right]^{\frac{1}{2}}\ln\left(\frac{h-d}{z_0}\right) \tag{14}$$

where C_{reg} is a regression constant that is dependent on soil and rooting depth and SW is the stem weight of the tree. See Gardiner et al. (2000) for more complete details.

Once the critical wind speeds have been calculated it is necessary to predict the likelihood of such a wind speed occurring at that location. The wind climate model used in the program is obtained from the DAMS scoring system. The DAMS score is found to be well correlated to the Weibull '*a*' parameter (Quine 2000) and the Weibull '*k*' parameter is assumed constant. The Weibull distribution is used to derive the extreme wind speed probability distribution (ESDU 1987) and hence the probability of occurrence of any wind speed. These probabilities are transformed into return periods for both overturning and breakage expressed in the average number of years likely to occur before damage.

Future probabilities of damage (Figure 5.17) are calculated with the aid of yield models (Edwards and Christie 1981). These allow the stands to grow in time so the program can estimate the annual probabilities for damage at different time steps. The temporal dimension of the model is particularly important as it allows estimation of the changing risk during the life of the crop, and for testing the best silviculture practices that may maintain the stability of the trees.

The first commercial release of the ForestGALES decision support system in 2000 was a purely non-spatial version. A second version has since been released which incorporates improved wind climatology, and a fully integrated GIS version of the model (Figure 5.18) is currently under development.

The GIS version will allow a visual analysis of the implications of silviculture strategies in terms of wind risk, such as thinning, retentions, design of felling coupes, new forest roads or the effect of clearfelling of neighbouring stands (edge effect).

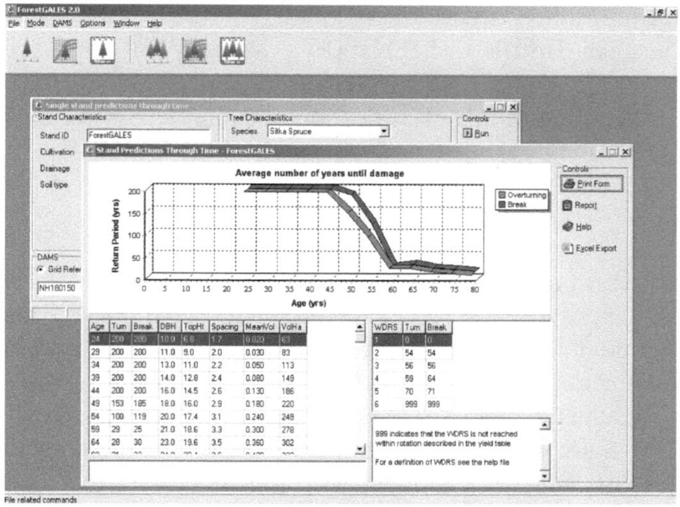

Figure 5-17. Example output screen from ForestGALES with the calculated return period displayed in the graph. Illustration courtesy of the Forestry Commission, UK.

Figure 5-18. The ForestGALES extension to ArcView GIS showing different levels of risk for part of Kielder forest in Northern England, UK. Illustration courtesy of the Forestry Commission, UK.

British Columbia System

The Canadian British Columbia (BC) Ministry of Forests diagnostic method is observational but includes some elements of the empirical approach. According to this classification, windthrow risk for an individual tree or a stand can be calculated as:

Windthrow Risk = Biophysical Risk + Treatment Risk

In this assessment, the **'Biophysical Risk'** is the combination of the topographic exposure, soil characteristics and stand hazard components representing the intrinsic windloading and wind stability of trees on the site prior to treatment. The **'Treatment Risk'** represents the way in which a particular treatment increases or decreases the windloading or wind resistance of trees, while the **'Windthrow Risk'** is a combination of the biophysical risk and the treatment risk and represents the likelihood of damage from endemic winds (Table 5-5).

Topographic exposure hazards are assessed on a large- and mid-scale, as well as on the base of the tree/stand position on the slope. This assessment is based on the principles of Alexander (1987).

Soil characteristics are included in the assessment since the strength of anchorage is a function of root-soil mass, root-soil bond or shallow soils and drainage. Trees with unrestricted root systems (in coarse alluvial/colluvial soils, with depth of rooting >0.8 m with good drainage) will have a low risk of windthrow, while root systems with impeded growth (in fine textured soils with rooting depth <0.4 m, impeded by high water table or impenetrable soil layer, with poor drainage) bear a high risk of windthrow.

Stand characteristics and exposure to prevailing winds are also assessed knowing that the risk increases with the mean tree height, H:D (stem taper) ratio, stand density, and the amount of inside-stand damage and decreases in multi-layered stands or in stands with high live crown ratio. Wide openings >5 tree lengths and those oriented downwind at right angles are most hazardous and upwind openings at right angles which are smaller than 2 tree lengths are of low risk. Commercial thinning of more than 50% of the basal area is considered as highly hazardous management strategy.

The first box grid of Table 5-5 integrates topographic exposure and soil risks which are intrinsic and constant, to yield 'Site Risk'. The site risk is integrated with stand risk, which changes as stands grow and management practices are applied. When brought together in the second box grid, they yield 'Overall Risk'. The results of the biophysical risk assessment should be checked in the field during the 'calibration' step and adjusted if necessary (Mitchell 1998).

Table 5-5. Diagnostic windthrow risk assessment method based on evaluation of the tree/stand topographic exposure, soil characteristics and stand characteristics (adapted from British Columbia Ministry of Forests 1999). L = Low, M = Moderate and H = High risk.

Site Risk		**Topographic Exposure**		
		Low	**Moderate**	**High**
Soils	Low	L	M	M
	Moderate	M	M	H
	High	M	H	H

Overall Risk		**Site Risk**		
		Low	**Moderate**	**High**
Stand	Low	L	M	M
	Moderate	M	M	H
	High	M	H	VH

BC Ministry of Forests recognises that the best practices against high windthrow risks should include:

- a statement of windthrow management objectives
- consideration of windthrow risk
- inclusion of strategies to minimize and recover windthrow
- identification and evaluation of windthrow risk
- integration of windthrow risk into choice of silvicultural system
- calculation of the **'Windthrow Impact'**, referring to the potential harm windthrow could cause if it occurs. The impact is negative if wind damage results in management objectives not being met. If some level of damage is acceptable, this should be indicated in the original silviculture prescription.

3.1.5 Tree stability under snow

In Europe, hundreds of millions of euros are lost annually because of snow and wind-associated damage to forests. The type of forest growing on a slope and its resistance to snow loading can also influence the likelihood and magnitude of avalanches occurring. Damage to single trees, and more importantly to forest stands, leads not only to losses of high-quality and high-value timber but also to detrimental insect attacks on the remaining stands and reduced seed production amongst the older trees. Unscheduled and costly thinnings are often a consequence of severe snow damage (Makinen and Isomaki 2004; Rochette et al. 2004; Seki et al. 2005; Tremblay and Begin 2005).

Snow accumulation on trees is highly dependent on the climatological and topographical conditions including:

- temperature: influences snow moisture content and, in turn, the degree to which it can stick to the branches and needles
- wind: causes the snow to be shed but also leads to large accumulations of wet snow (late autumn or early spring), rime, or freezing rain
- geographic location and topography: affect the occurrence of damaging forms of snow, e.g., coastal locations and moderate to high elevations usually get large snow accumulations
- slope angle and aspect play a less important role but the evidence on the role of aspect is contradictory.

The severity of snow damage is related to tree characteristics that control the overall stability:

- stem taper and crown characteristics: slightly tapering stems, asymmetric crowns and rigid horizontal branching are highly hazardous
- species: due to coupling with the specific location, the hazard of failure for a particular species can not be clearly defined
- stand and forest management: can alter the hazards posed by the snow through choice of regeneration, tending, thinning and rotation.

For more information regarding the stability of trees under snow, the reader is referred to the following texts: Paatalo et al. (1999); Paatalo (2000); Peltola et al. (1997, 1999, 2000).

4. REFERENCES

Abe K, Iwamoto M (1986) Preliminary experiment on shear in soil layers with a large direct shear apparatus. J Jpn For Soc 68:61-65

Alexander RR (1987) Ecology, silviculture and management of Engelmann spruce and subalpine fir type in central and southern Rocky Mountains. USDA For Ser Agric Handbook No. 659

Ancelin P, Courbaud B, Fourcaud T (2004) Development of an individual tree-based mechanical model to predict wind damage within forest stands. For Ecol Manage 203:101-121

Barker DH (1987) A3.2.9 Rooting effects. In: Hewlett HWM, Boorman LA, Bramley ME (eds) Design of reinforced grass waterways. Report 116, CIRIA, London

Biddle PG (1998) Tree Root Damage to Buildings. Willowmead Publishing, Wantage

Bishop AW (1955) The use of the slip circle in the stability analysis of earth slopes. Geotechnique 5:7-17

Blackburn P, Petty JA (1988) Theoretical calculations of the influence of spacing on stand stability. Forestry 61:29-43

Bowyer-Bower TAS, Burt TP (1989) Rainfall simulators for investigating soil response to rainfall. Soil Technol 2:1-16

Brang P (2001) Resistance and elasticity: promising concepts for management of the protection forests in the European Alps. For Ecol Manage 145:107-119

Brinkgreve RBJ (2002) Plaxis 2D – Version 8 Manual. Balkema, Lisse

British Columbia Ministry of Forests (1999) Mapping and Assessing Terrain Stability Guidebook. 2nd edn. For. Prac. Br., Victoria, BC

Brown CB, Sheu MS (1975) Effects of deforestation on slopes. J Geotech Eng – ASCE, 101(GT2):147-165

BS6031 (1981) Code of practice for earthworks. HMSO, London

BS EN 1997-1 (2004) Eurocode 7. Geotechnical Design. General Design.

BS EN 1997-2 (2007) Eurocode 7. Geotechnical Design. Ground Investigation and Testing.

Buchanan P, Savigny KW (1990) Factors controlling debris avalanche initiation. Can Geotech J 27:659-675

Burroughs ER, Thomas BR (1977) Declining root strength in Douglas-fir after felling as a factor in slope stability. USDA Forest Service Research Paper INT-190, 1-27

Busby JA (1965) Studies on the stability of conifer stands. Scot Forest 19:86-102

Cammeraat LH (2004) Scale dependent thresholds in hydrological and erosion response of a semi-arid catchment in Southeast Spain. Agr Ecosys Environ 104:317-332

Casenave A, Valentin C (1989) Les états de surface de la zone Sahélienne. Influence sur l'infiltration. ORSTOM, Collection Didactiques, Paris, pp 1-230

Cazzuffi D, Corneo A, Crippa E (2006) Slope stabilisation in Southern Italy: plant growth and temporal performance. Geotech Geol Eng 24:429-447

Cooke R, Warren A, Goudie A (1993) Desert Geomorphology. UCL Press, London

Coppin NJ and Richards IJ (1990) Use of Vegetation in Civil Engineering. CIRIA, Butterworths, London

Cremer KW, Borough CJ, McKinnel FH, Carter PP (1982) Effects of stocking and thinning on wind damage in plantations. N Zeal J For Sci 12:245-268

Cucchi V, Meredieu C, Stokes A, de Coligny F, Suárez J, Gardiner B (2005) Modelling the windthrow risk for simulated forest stands of maritime pine (*Pinus pinaster* Ait.). For Ecol Manage 213:184-196

Danjon F, Barker DH, Drexhage M, Stokes A (2007) Using 3D plant root architecture in models of shallow slope stability. Ann Bot-London, in press

De Champs J (1987) Mesures sylvicoles préventives. Revue Forestière Française 39:313-322

Duncan JM, Wright SG (2005) Soil Strength and Slope Stability. Wiley and Sons, Inc., New Jersey

Dvorak L, Bachmann P, Mandallaz D (2001) Sturmschaden in ungleichformigen bestanden. Schweiz Z Forstwes 152

Edwards PN, Christie JM (1981) Yield models for forest management. Forestry Commission Booklet No. 48. Forestry Commission, Edinburgh

Ekanayake JC, Phillips CJ (1999a) A model for determining thresholds for initiation of shallow landslides under near-saturated conditions in the East Coast region, New Zealand. J Hydrol (NZ) 38(1):1-28

Ekanayake JC, Phillips CJ (1999b) A method for stability analysis of vegetated hillslopes: an energy approach. Can Geotech J 36:1172-1184

Ekanayake JC, Phillips CJ (2002) Slope stability thresholds for vegetated hillslopes: a composite model. Can Geotech J 39:849-862

Ekanayake JC, Marden M, Watson AJ, Rowan D (1997) Tree roots and slope stability: a comparison between *Pinus radiata* and kanuka. New Zeal J For Sci 27:216-233

Endo T, Tsuruta T (1969) On the effect of tree roots upon the shearing strength of soil. Annual report of the Hokkaido Branch, Forest Place Experimental Station, Sapporo, Japan, pp 167-183

ESDU (1987) World-wide extreme wind speeds. Part 1: Origins and methods of analysis. Data Item 87034, ESDU International, London, p 47

Fellenius W (1936) Calculations of the stability of earth dams. In: Trans. 2nd Congress on Large Dams, Washington, vol 4, p 445

Fredlund DG, Rahardjo H (1993) Soil Mechanics for Unsaturated Soils. Wiley and Sons, New York

Galinski W (1989) A windthrow-risk estimation for coniferous trees. Forestry 62:139-146

Gardiner BA, Quine CP (2000) Management of forests to reduce the risk of abiotic damage – a review with particular reference to the effects of strong winds. For Ecol Manage 135: 261-277

Gardiner BA, Stacey GR, Belcher RE, Wood CJ (1997) Field and wind tunnel assessments of the implications of respacing and thinning for tree stability. Forestry 70:233-252

Gardiner BA, Peltola H, Kellomäki S (2000) Comparison of two methods for predicting the critical wind speed required to damage coniferous trees. Ecol Model 129:1-23

Gardiner B, Suarez J, Achim A, Hale S, Nicoll B (2004) ForestGALES. A PC-based wind risk model for British forests. Version 2.0. Forestry Commission, Edinburgh

Gardiner BA, Marshall B, Achim A, Belcher R, Wood C (2005) The stability of different silvicultural systems: a wind-tunnel investigation. Forestry 78:471-484

Glade T, Anderson M, Crozier MJ (2005) Landslide Hazard and Risk. Wiley & Sons, Chichester

Goodman RE (1999) Karl Terzaghi: The Engineer as Artist. ASCE Press. pp 340

Gray DH (1995) Keynote address: Influence of vegetation on the stability of slopes. Proceedings of the International Conference on Vegetation and Slopes, Stabilisation, Protection and Ecology, University Museum, Oxford, 29-30 September 1994, Thomas Telford, London, pp 1-24

Gray DH, Megahan WF (1981) Forest vegetation removal and slope stability in the Idaho batholith, United States Department of Agriculture Forest Service, Intermountain Forest and Range Experimental Station Research Paper, INT-271:1-23

Greenwood JR (1983) A simple approach to slope stability. Ground Eng 16:45-48

Greenwood JR (1987) Effective stress stability analysis. In: 9th European Conference on Soil Mechanics and Foundations, Dublin, September 1987, Balkema, vol 3, pp 1082-1083

Greenwood JR (1989) Design approach for slope repairs and embankment widening. Reinforced Embankments Symposium, Cambridge September 1989. Thomas Telford Ltd, pp 51-61

Greenwood JR (1990) Inclusion of reinforcement forces in stability analysis. Geotextiles, Geomembranes and Related Products 114:997-999

Greenwood JR (2006) Slip4ex – A program for routine slope stability analysis to include the effects of vegetation, reinforcement and hydrological changes. Geotech Geol Eng 24:449-465

Greenwood JR, Vickers AW, Morgan RPC, Coppin NJ, Norris JE (2001) Bioengineering – the Longham Wood Cutting field trial. CIRIA PR 81, London

Greenwood JR, Norris JE, Wint J, Barker DH (2003) Bioengineering and the transportation infrastructure. In: Frost MW, Jefferson I, Faragher E, Roff TEJ, Fleming PR (eds) Transportation Geotechnics. Thomas Telford, London, pp 205-220

Greenwood JR, Norris JE, Wint J (2004) Assessing the contribution of vegetation to slope stability. J Geotech Eng 157:199-208

Haigh MJ (1977) The use of erosion pins in the study of slope evolution. In: Shorter Technical Methods (II), Technical Bulletin No 18, British Geomorphological Research Group, Geo Books, Norwich, UK

Hill J, Schütt B (2000) The use of remote sensing satellites for mapping complex patterns of erosion and stability in arid Mediterranean ecosystems. Remote Sens Environ 74: 557-569

Hsi G, Nath JH (1970) Wind drag within a simulated forest. J Appl Meteorol 9:592-602

Hudson NW (1979) Soil Conservation. Batsford Lim. London

Hudson NW (1993) Field measurement of soil erosion and runoff, FAO, Rome; online on: http://www.fao.org/documents/show_cdr.asp?url_file=/docrep/T0848E/t0848e00.htm

Hunt R, Dyer RH, Driscoll R (1991) Foundation movement and remedial underpinning. BRE Report 184

Imeson AC, Cammeraat LH (1999) Scaling up from field measurements to large areas using the Desertification Response Unit and Indicator Approaches. In: Arnalds O, Archer S (eds), Rangeland Desertification. Advances in Vegetation Science 19. Kluwer Academic Publishers, Dordrecht, pp 99-114

Indraratna B, Fatahi B, Khabbaz H (2005) Numerical analysis of matric suction effects induced by tree roots. Geotech Eng 159:77-90

Itasca (2002) FLAC 4.0 User Manual

Janbu N (1973) Slope stability computations. In: Hirschfield RC, Poulos SJ (eds) Embankment Dam Engineering. Wiley, New York

Keim RF, Skaugset AE (2003) Modelling effects of forest canopies on slope stability. Hydrol Proc 17: 1457-1467

Krahn J (2001) The R. M. Hardy Keynote Address: The limits of limit equilibrium analysis. Can Geotech J 40:643-660

Lanier L (1994) Précis de sylviculture. ENGREF, Nancy

Lascelles BDT, Favis-Mortlock DT, Parsons AJ, Guerra AJT (2000) Spatial and temporal variation in two rainfall simulators: implications for spatially explicit rainfall simulation experiments. Earth Surf Proc Land 25:709-721

Lohmander P, Helles F (1987) Windthrow probability as a function of stand characteristics and shelter. Scand J For Res 2:227-238

Maccurrach RS (1991) Spacing: an option for reducing storm damage. Scot Forest 45:285

Makinen H, Isomaki A (2004) Thinning intensity and long-term changes in increment and stem form of Norway spruce trees. For Ecol Manage 201:295-309

Mason WL (2002) Are irregular stands more windfirm? Forestry 75:347-355

Mattia C, Bischetti GB, Gentile F (2005) Biotechnical characteristics of root systems of typical Mediterranean species. Plant Soil 278:23-32

Metternicht GI, Fermont A (1998) Estimating erosion surface features by linear mixture modeling – contribution of remote sensing, a review. Remote Sens Environ 64:254-265

Mickovski SB, van Beek LPH (2006) A decision support system for the evaluation of eco-engineering strategies for slope protection. Geotech Geol Eng 24:483-498

Mickovski SB, Stokes A, van Beek LPH (2005) A decision support tool for windthrow hazard assessment and prevention. For Ecol Manage 216:64-76

Mickovski SB, Bengough AG, Bransby MF, Davies MCR, Hallett PD, Sonnenberg R (2007a) Material stiffness, branching pattern and soil matric potential affect the pullout resistance of model root systems. Eur J Soil Sci, doi: 10.1111/j.1365-2389.2007.00953

Mickovski SB, Sonnenberg R, Bransby MF, Davies MCR, Lauder K, Bengough AB, Hallett PD (2007b) Shear reinforcement of soil by vegetation. Proceedings of the Fourteenth European Conference on Soil Mechanics and Geotechnical Engineering, Madrid 24-27 September 2007. Millpress Science Publishers, Rotterdam, The Netherlands, pp 798-783

Miller KF (1985) Windthrow hazard classification. Forestry Commission Leaflet 85, HMSO, London

Miller KF, Quine CP, Hunt J (1987) The assessment of wind exposure for forestry in upland Britain. Forestry 60 (2): 179-192

Mitchell SJ (1998) A diagnostic framework for windthrow risk estimation. For Chron 74: 100-105

Morgan RPC (2005) Soil Erosion and Conservation. 3rd edn. Blackwell Publishing, Oxford

Morrison IM, Greenwood JR (1989) Assumptions in simplified slope stability analysis by the method of slices. Geotechnique 39:503-509

Nicoll BC, Ray D (1996) Adaptive growth of tree root systems in response to wind action and site conditions. Tree Physiol 16:891-898

Nicoll BC, Achim A, Mochan S, Gardiner BA (2005) Does steep terrain influence tree stability? – A field investigation. Can J Forest Res 35:2360-2367

Nicoll BC, Gardiner BA, Rayner B, Peace AJ (2006) Anchorage of coniferous trees in relation to species, soil type and rooting depth. Can J Forest Res 36:1871-1883

Norris JE (2005a) Root mechanics applied to slope stability. PhD thesis, Nottingham Trent University, Nottingham, UK

Norris JE (2005b) Root reinforcement by hawthorn and oak roots on a highway cut-slope in Southern England. Plant Soil 278:43-53

Norris JE, Greenwood JR (2000) In situ shear and pull out testing to demonstrate the enhanced shear strength of root reinforced soil. In: Proceedings of the 8th International Symposium on Landslides, Cardiff, 26-30 June 2000. Thomas Telford, London, pp 1123-1128

Norris JE, Greenwood JR (2003) In-situ shear box and root pull-out apparatus for measuring the reinforcing effects of vegetation. In: Myrvoll F (ed) Field Measurements in Geomechanics, Oslo. Swets and Zeitlinger, Lisse, pp 593-597

Norris JE, Greenwood JR (2006) Assessing the role of vegetation on soil slopes in urban areas. IAEG2006, Geological Society of London, Paper no. 744, 1-12

O'Loughlin CL (1974) The effect of timber removal on the stability of forest soils. J Hydrol (N Z) 13(2):121-34

O'Loughlin CL, Ziemer RR (1982) The importance of root strength and deterioration rates upon edaphic stability in steepland forests. In: Warring RH (ed) Carbon uptake and allocation in subalpine ecosystems as a key to management. Proceedings of an I.U.F.R.O. workshop P.I. 107-00 Ecology of subalpine zones, August 2-3 Oregon State University, Corvallis, Oregon, USA, pp 70-78

O'Loughlin CL, Rowe LK, Pearce AJ (1982) Exceptional storm influences on slope erosion and sediment yield in small forest catchments, North Westland, New Zealand. In O'Loughlin EM, Brens LJ (eds) First National Symposium on Forest Hydrology, Melbourne, 1982. Institution of Engineers, Australia, National Conference Publication 82/6, pp 84-91

Operstein V, Frydman S (2000) The influence of vegetation on soil strength. Ground Improv 4(2):81-89

Otto H-J (2000) Expériences sylvicoles après des ouragans catastrophiques: regard dans le passé en Basse-Saxe. Revue Forestière Française 52:223-238

Paatalo ML, Peltola H, Kellomaki S (1999) Modelling the risk of snow damage to forests under short-term snow loading. For Ecol Manage 116:51-70

Paatalo ML (2000) Risk of snow damage in unmanaged and managed stands of Scots pine, Norway spruce and birch. Scan J For Res 15:530-541

Peltola H, Kellomaki S (1993) A mechanistic model for calculating windthrow and stem breakage of Scots pines at stand edge. Silva Fenn 27:99-111

Peltola H, Nykanen ML, Kellomaki S (1997) Model computations on the critical combination of snow loading and windspeed for snow damage of Scots pine, Norway spruce and Birch sp. at stand edge. For Ecol Manage 95:229-241

Peltola H, Kellomaki S, Vaisanen H, Ikonen VP (1999) A mechanistic model for assessing the risk of wind and snow damage to single trees and stands of Scots pine, Norway spruce and birch. Can J For Res 29:647-661

Peltola H, Gardiner B, Kellomaki S, Kolstrom T, Lassig R, Moore J, Quine C, Ruel JC (2000) Wind and other abiotic risks to forests – Introduction. For Ecol Manage 135:1-2

Quine CP (2000) Estimation of mean wind climate and probability of strong winds for wind risk assessment. Forestry 73:247-258

Quine CP (2002) The role of wind in the ecology and naturalisation of Sitka spruce in upland Britain. PhD thesis, University of Edinburgh

Quine CP, White IMS (1993) Revised windiness scores for the windthrow hazard classification: the revised scoring method. Forestry Commission, Farnham

Quine CP, Gardiner BA (1998) ForestGALES – Replacing the windthrow hazard classification. In: Forest Research Report and Accounts 1997-98. The Stationery Office, Edinburgh, pp 27-31

Quine CP, Coutts MP, Gardiner BA, Pyatt DG (1995) Forests and Wind: Management to minimise damage. Forestry Commission Bulletin 114. HMSO, London

Quine TA, Walling DE, Chakela QK, Mandiringana OT, Zhang X (1999) Rates and patterns of tillage and water erosion on terraces and contour strips: evidence from caesium-137 measurements. Catena 36:115-142

Riestenberg MM, Sovonick-Dunford S (1983) The role of woody vegetation in stabilising slopes in the Cincinnati area. Geol Soc Am Bull 94:504-518

Rochette P, Belanger G, Castonguay Y, Bootsma A, Mongrain D (2004) Climate change and winter damage to fruit trees in eastern Canada. Can J Plant Sci 84:1113-1125

Savill PS (1983) Silviculture in windy climates. Forestry Abstracts 44:473-488

Schofield AN (1998) Mohr Coulomb error correction. Ground Eng August:30-32

Schofield AN (1999) A note on Taylor's interlocking and Terzaghi's "true cohesion" error. Geotechnical News 17:4

Schmidt KM, Roering JJ, Stock JD, Dietrich WE, Montgomery DR, Schaub T (2001) The variability of root cohesion as an influence on shallow landslide susceptibility in the Oregon Coast range. Can Geotech J 38:995-1024

Schütz J-P (1997) Sylviculture 2: la gestion des forêts irrégulières et mélangées. Presses Polytechniques et Universitaires Romandes, Lausanne, p 178

Seki T, Kajimoto T, Sugita H, Daimaru H, Ikeda S, Okamoto T (2005) Mechanical damage on *Abies mariesii* trees buried below the snowpack. Arc Ant Alpine Res 37:34-40

Swanston DN (1970) Mechanics of debris avalanching in shallow till soils of southeast Alaska. U.S. Forest Service, Research Paper PNW-103, Pacific and Northwest Forest and Range Experimental Station, Portland, Oregon

Takken I, Govers G, Jetten V, Nachtergaele J, Steegen A, Poesen J (2001) Effects of tillage on runoff and erosion patterns. Soil Till Res 61:55-60

Tobias S (1995) Shear strength of the soil root bond system. In: Barker DH (ed) Proceedings of the International Conference on Vegetation and Slopes, Stabilisation, Protection and Ecology, University Museum, Oxford, 29-30 September 1994. Thomas Telford, London. pp 280-286

Tongway D (1994) Rangeland soil condition assessment manual. CSIRO, Division of Wildlife and Ecology Canberra, pp 1-69

Tongway D, Hindley N (1995) Manual for assessment of soil condition tropical grasslands. CSIRO, Division of Wildlife and Ecology, Canberra, pp 1-60

Tremblay J, Begin Y (2005) The effects of snow packing on tree growth forms on an island in a recently created reservoir in northern Quebec, Canada. Ecoscience 12:530-539

Troen I, Petersen EL (1989) European Wind Atlas. Published for the Commission of the European Communities by Risø National Laboratory, Roskilde, Denmark, pp 656

Turnbull WJ, Hvorslev MJ (1967) Special problems in slope stability. J Soil Mech Eng ASCE 93(SM4):499-528

Valentin C, Bresson LM (1992) Morphology, genesis and classification of surface crusts in loamy and sandy soils. Geoderma 55:225-245

Valinger E, Lundqvist L, Bondesson L (1993) Assessing the risk of snow and wind damage from tree physical characteristics. Forestry 66:249-260

Van Beek LPH, Wint J, Cammeraat LH, Edwards P (2005) Observation and simulation of root reinforcement on abandoned Mediterranean slopes. Plant Soil 278:55-74

Vandaele K, Vanommelslaeghe J, Muylaert RAF, Govers G (1996) Monitoring soil redistribution patterns using sequential aerial photographs. Earth Surf Proc Land 21:353-364

Verstraeten G, Poesen J (2000) Estimating trap efficiency of small reservoirs and ponds: methods and implications for the assessment of sediment yield. Prog Phys Geog 24:219-252

Waldron LJ, Dakessian S (1981) Soil reinforcement by roots: calculation of increased soil shear resistance from root properties. Soil Sci 132:427-435

Waldron LJ, Dakessian S (1982) Effect of grass, legume, and tree roots on soil shearing resistance. J Soil Sci Soc Am 46:894-899

Waldron LJ, Dakessian S, Nemson JA (1983) Shear resistance enhancement of 1.22-meter diameter soil cross sections by pine and alfalfa roots. J Soil Sci Soc Am 47:9-14

Walling DE, Quine TA (1990) The use of Caesium 137 to investigate patterns and rates of soil erosion on arable fields. In: Boardman J, Foster IDL, Dearing JA (eds) Soil Erosion in Agricultural Land. Wiley and Sons, Chichester, pp 33-53

Wischmeier WH (1978) Use and misuse of the Universal Soil Loss Equation. J Soil Water Con 31:5-9

Wischmeier WH, Smith DD (1978) Predicting rainfall erosion losses-a guide to conservation planning. Agriculture Handbook 537, U.S. Department of Agriculture, Washington D.C.

Wu TH (1984a) Effect of vegetation on slope stability. In: Soil reinforcement and moisture effects on slope stability. Transportation Research Board, Washington, D.C. pp 37-46

Wu TH (1984b) Soil movements on permafrost slopes near Fairbanks, Alaska. Can Geotech J 21:699-709

Wu TH, McKinnell III WP, Swanston DN (1979) Strength of tree roots and landslides on Prince of Wales Island, Alaska. Can Geotech J 16:19-33

Yang B, Raupach MR, Shaw RH, U KTP, Morse AP (2006) Large-eddy simulation of turbulent flow across a forest edge. Part I: Flow statistics. Bound-Lay Meteorol 120:377-412

Ziemer RR (1981) Roots and shallow stability of forested slopes. Int Ass Hydrol Sci 132:343-361

Ziemer RR, Swanston DN (1977) Root strength changes after logging in southeast Alaska. USDA Forest Service, Research Note PNW-306. Forest Service, USDA, Portland, pp 10

Chapter 6

SPECIES SELECTION FOR SOIL REINFORCEMENT AND PROTECTION

J.E. Norris[1,2], A. Di Iorio[3], A. Stokes[4], B.C. Nicoll[5], A. Achim[6]

[1] Halcrow Group Limited, Endeavour House, Forder Way, Cygnet Park, Hampton, Peterborough, PE7 8GX, U.K., [2] School of Architecture, Design and Built Environment, Nottingham Trent University, Burton Street, Nottingham, NG1 4BU, U.K., [3]Dept. Chemical and Environmental Sciences, University of Insubria, Via Valleggio 11, CO-22100, Italy, [4] INRA, AMAP, A A-51/PS2, Boulevard de la Lironde, 34398 Montpellier cedex 5, France, [5] Forest Research, Northern Research Station, Roslin, Midlothian, EH25 9SY, U.K., [6] Faculté de Foresterie et de Géomatique, Université Laval, Québec, G1K 7P4, Canada

Abstract: *Species selection is vitally important for ensuring the success of any eco-technological solution that may be employed on a particular site. The purpose of this chapter is to provide the engineer with a database of plant species that are suitable for both soil and slope stability by either mechanical or hydrological means, i.e., anchoring and buttressing of deep tap roots; bank and channel reinforcement; deep reinforcement and soil strength enhancement; removing soil moisture, surface protection, shallow reinforcement and erosion control. Protection forests rely on the stability of trees to maintain their integrity especially during storms and with regard to rockfall or avalanches. We therefore provide guidelines as to which species best resist these abiotic forces on slopes.*

Key words: grasses, pioneer plants, plant morphology, role of vegetation, shrubs, soil re-inforcement, trees

1. INTRODUCTION

Plants can fulfil many functional roles, therefore selection of the right species for a particular purpose is essential in ensuring the efficacy of the ecotechnological solution. Selecting native plants will usually increase the success of the planting program and reduce the long-term maintenance requirements. However, in plant ecology, it is well known that vegetation

J.E. Norris et al. (eds.), Slope stability and erosion control: Ecotechnological solutions, 167–210.

tends to use all the attainable resources but with low efficiency. As a consequence, the first period, especially during the first year, following planting or afforestation practices is crucial for a successful field performance. Providing regular monitoring and maintenance also assures adequate plant survival.

The suitability of a plant for ground bio-engineering measures depends on the characteristics, requirements and structure of the plant, its usability for certain building systems (see Chapter 7) and its resistance to mechanical forces caused by any form of soil erosion and instability. Species used for conservation of soil properties and promoting soil stabilisation should meet the following criteria:

1. pioneer plants which grow rapidly on degraded land, landslides, gullies and new road slopes;
2. dense and deep rooting systems which add strength to the surface soil layers and increase the shear strength;
3. adventitious rooting ability and coverage resistance;
4. fast and simple propagation such as cuttings and their application in the dormant season (Weigel et al. 1987; Lammeraaner et al. 2005).

The chosen species must also fulfill environmental and practical requirements such as those given in Table 6-1 and in the following list (Lammeranner et al. 2005):

1. range of altitude;
2. hill slope aspect;
3. moisture, light and soil requirements;
4. economic value for local population;
5. preferences of local population;
6. availability of species in local nurseries;
7. planting condition, size of mature plant, form and habit, and recommended spacings.

In this chapter, tables are included which list the plants that can be used for ecotechnological solutions and the function of that plant e.g. soil stabilization, erosion control and protection against rockfall (Tables 6-2, 6-3, 6-4). The tables provide information regarding region, habitat, soil properties and the tolerance of the species to drought, flooding, storms, etc, altitude ranges, morphology of the plant above and below ground, whether the species is a pioneer plant and the role of the species in protecting soil stability.

More detailed information on plant selection for slope stability and soil erosion are provided in Sections 1.1 and 1.2. The recent concept of using appropriate vegetation in zones of severe erosion, or "hotspots" is presented along with perspectives for future research.

Table 6-1. Factors to take into consideration when selecting the best plants for erosion control or slope stabilisation (Myers 1993).

Water Requirements	**Light Requirements**	**Rooting Characteristics**	**Planting Condition**	**Comments**
Dry - Once established, tolerates dry soil conditions during the growing season	**Full Sun** - Requires sun throughout the day	**Fibrous** - Lacks a central root; root mass composed of fibrous lateral roots	Sizes given are those that are generally found in nurseries; other sizes may also be available	Growth rate; ornamental and wildlife value; wind/salt spray tolerance; maintenance; average life span
Moist - Requires moist soil throughout the growing season	**Sun/Shade** - Requires shade for about 1/2 the day	**Tap** - With a stout, central main root		Indigenous species preferred to exotic species
Wet - Tolerates saturated soil year-round	**Full Shade** - Requires shade throughout the day	**Shallow, Moderate, Deep** refers to relative rooting depth (influenced by soil and groundwater conditions		Consider carefully problem on site and long-term consequences, in particular with regard to geomorphological hazards e.g., rockfall and avalanches
Usage - Relative water uptake by plant [e.g., high or no data]				

1.1 Plant selection for slope stability

1.1.1 Grasses and herbs

It is recognised that the contribution of several herbaceous and shrub species to slope stability is largely indirect, i.e. hydraulically, rather than direct, i.e. mechanically. In natural conditions, the colonisation of bare stream banks and forest sites by herbaceous vegetation in the post-harvesting or landslide phases is a consequence of tree canopy removal admitting light for establishment of opportunistic (pioneer) species. Vegetation protects the surface directly from rain splash (and throughfall drip below a forest canopy) and the roots and rhizomes help to bind the soil (Gyssels et al. 2005; Bochet et al. 2006), for example, study of rainfall interception in a mixed-grass prairie characterised by the Agropyron-Koeleria association (Couturier and Ripley 1973), reported that in the ungrazed area with plants below 0.15 m in height, an interception value rose from 21 – 32% during two years of measurements. In an adjoining grazed area, interception was 70% lower.

Figure 6-1. An example of incipient earthflow showing displacement under an unruptured membrane of pasture vegetation, but an absence of surface translation (from Preston and Crozier 1999, reprinted by permission of the publisher).

A further example of how grasses contribute to soil stabilization can be seen in Figure 6.1. An example of a regolith unit is shown, which has experienced internal deformation and possible fluidization, but lateral translation has not occurred. The absence of translational movement characteristic of this phenomenon is attributed to the constraining influence of a surface membrane of densely interwoven roots of pasture species, grasses and/or forbs (Preston and Crozier 1999).

A statistical evaluation of factors affecting alpine slope stability has also shown that land use is an important factor to consider when evaluating landslides in topsoils. Tasser et al. (2003) found that managed meadows and pastures were less erodible than abandoned grasslands. However, the land-use activities themselves did not lead to changes in erosion risks, but rather had direct or indirect effects on vegetation and soil properties. Changes incurred included a decrease in the relative cover of grasses, herbs and dwarf shrubs as well as the total root length and the rooting density in main fracture depth (Tasser et al. 2003). In abandoned pastures, tussock grasses (e.g. *Nardus stricta* L. and *Dactylis glomerata* L.) and tall, rigid dwarf shrubs (e.g. *Vaccinium myrtillus* L. and *Vaccinium vitis-idea* L.) can be abundant and such species are resistant to snow gliding in winter. Therefore, downward forces of sliding snow on the frozen plant parts may result in tension fissures in the soil, ultimately leading to landslides (Tasser et al. 2003). Nevertheless, the slowness and irregularity of passive restoration in alpine environments is well-known (see Muller et al. 1998) but if adapted species were sown (Schmid et al. 2007), this could lead to the constitution of an artificial ecosystem, which can be progressively replaced by native plant communities.

In a study concerning the evolution of artificially sown alpine meadows, three main stages in the evolution of these communities occurred (Bédécarrats 1991). During the first stage (1st to 3rd year), the sown species and some ruderals[1] dominated. The grassland seemed artificial compared to the neighbouring spontaneous vegetation. The second stage was transitional (4 – 5th year), corresponding to colonisation by native species such as *Dactylis glomerata*, *Rumex acetosella* L., *Trifolium pratense* L., *Deschampsia caespitosa* (L.) Beauv., *Achillea millefolium* L., and *Tussilago farfara* L. at the subalpine level and *Poa alpina* L. and *Alchemilla vulgaris* L. at the alpine level. As a consequence, species richness increased as a whole while the contribution of sown species decreased. Finally, the third stage (6th year and after) was a maturation stage during which mid-succession species appeared, including *Hypericum maculatum* Crantz, *Epilobium angustifolium* L., and *Leucanthemum vulgare* Lam. at low altitudes and *Festuca violacea* Schleicher ex Gaudin, *Carex sempervirens* Vill. and *Alchemilla pentaphyllea* L. at higher altitudes. Average species richness rose from 20 species in the sixth year to more than 30 in the ninth year. The communities then entered a phase of stabilisation and resembled spontaneous grasslands at similar altitudes.

[1] Plants that grow on poor land and waste ground.

Therefore, the success of the recolonisation processes depends on the maintenance of species-rich agricultural meadows and also on the marginal grassland plant habitats such as road verges, edges, plot boundaries, or even hedgerows (Alard et al. 1994). The ecological networks constituted by patches and corridors improve connectivity between seed reservoirs and restoration plots (Forman and Godron 1986). Restoring species-rich grasslands will depend on the maintenance of landscape diversity by agricultural areas where mainly rejuvenation procedures have to be performed. In contrast, grassland restoration in areas of intensive agriculture or after civil engineering projects in large-scale denuded landscapes appears to be a long-term process.

1.1.2 Shrubs and trees

When trees are planted on slopes, they are not only susceptible to instability from landslides and mass movements but also from extreme climatic conditions, such as storms. In the case of landslides or slope instability, trees will fail depending partly on the form of their root systems (Wu 2007; see Chapter 4). Under an increasing load acting along the slope axis, the action of the involved forces is counteracted directly by a stiff taproot or vertical roots (e.g. *Quercus* sp., *Pinus* sp.) meaning that a tap-rooted tree is more resistant than a tree with a heart- or shallow plate-root system (e.g. *Picea* sp.). Tree species with these three types of root systems fail differently and during a landslide, a taprooted tree will more likely develop the full tensile strength of the taproot whereas in plate or heart rooted trees, many roots exist which cross the slip surface at different angles and therefore do not fail in tension at large shear displacements (Wu et al. 2004). With regard to resistance to windthrow however, heart- and tap-rooted systems are generally more mechanically stable than plate root systems (see Chapter 4). Certain species have been classed in order of resistance to windthrow in Table 6-2, but care must be taken with this classification, which is highly site dependent (Bouchon 1987). Not only should species be windfirm and economically viable, but well adapted to the site in terms of water and nutrient availability as well as climatic conditions.

Table 6-2. Species resistance to windthrow. Care must be taken when using this table, which is indicative only.

Most resistant			→				Least resistant	
Oak	Beech	Larch	Douglas fir	Pine	Birch	Fir	Poplar	Spruce

A list of tree species with different types of root systems is given in Table 6-3, however, this list is only indicative, as most root systems are highly plastic i.e. root architecture is changed significantly, depending on local soil conditions. With regard to shrubby species, it is much more difficult to classify root architecture and little information exists on this subject. Most detailed studies on root architecture of shrubs have been carried out by Kutschera and Lichtenegger (1997, 2002) and the reader is referred to these books and their excellent drawings of root systems of many European species.

Table 6-3. Root system types which can be commonly found in different forest tree species. Species in brackets are highly plastic, i.e. root system architecture changes depending on local soil conditions. From Stokes (2002) using data compiled from Bűsgen et al. (1929), Kőstler et al. (1968), Eis (1978), Kutschera and Lichtenegger (1997, 2002) and Wu (2007).

Type of root system		
Plate	**Heart**	**Tap**
(*Betula pendula* Roth.)	*Acer campestre* L.	*Abies alba* Mill.
Fraxinus excelsior L.	*Acer platanoides* L.	*Juniperus communis* L.
Picea abies L.	*Acer pseudoplatanus* L.	(*Quercus sp.*)
Picea sitchensis Bong.	*Alnus glutinosa* L.	*Pinus contorta* Dougl.
Pinus cembra L.	*Alnus incana* L.	*Pinus nigra* Arnold
Pinus radiata D.	*Betula verrucosa* Ehrh.	*Pinus pinaster* Ait.
Pinus strobus L.	*Carpinus betulus* L.	*Pinus sylvestris* L.
(*Populus sp.*)	*Crateagus monogyna* Jacq.	*Pyrus pyraster* Burgsd.
Populus tremula L.	*Castanea sativa* Mill.	(*Robinia pseudoacacia* L.)
(*Robinia pseudoacacia* L.)	(*Fagus sylvatica* L.)	*Sorbus torminalis* L.
(*Sorbus aucuparia* L.)	*Larix decidua* Mill.	
	Larix leptolepis Sieb.	
	(*Populus sp.*)	
	Prunus avium L.	
	Pseudotsuga menziesii Mirb.	
	Pseudotsuga taxifolia Britt.	
	Quercus petraea Liebl.	
	Quercus robur L.	
	Quercus rubra L.	
	Taxus baccata L.	
	Tilia cordata Mill.	
	Tilia platyphyllos Scop.	
	Ulmus effusa Willd.	
	Ulmus glabra Huds.	
	Ulmus montana With.	

Root system morphology (see Chapter 4) is thus controlled both genetically and by environmental conditions. The development of any particular root architecture in response to either of these factors dictates its contribution to slope stability. In general, root systems with strong, deeply penetrating vertical or sinker roots that penetrate potential shear surfaces are more likely to increase stability against shallow sliding. A high density or concentration of small diameter fibrous roots is also more effective than a few large diameter roots for increasing the shear strength of a root-permeated soil mass. Roots must penetrate across the potential shear surface to have a significant effect Figure 6.2 (Cammeraat et al. 2005; van Beek et al. 2005). The most effective reinforcement is provided where roots penetrate across the soil mantle into fractures of fissures in the underlying bedrock or where roots penetrate into a residual soil or transition zone. When these conditions are present, density and shear strength will increase with depth.

Several studies showed that with regard to number and in biomass, 80 – 90% of tree roots are concentrated in the upper 0.9 m of soil (Tsukamoto and Kusakabe 1984; Gray and Sotir 1996; Di Iorio et al. 2005). The bulk of the near-surface roots are laterals; in contrast, roots below 0.9 m are generally oriented vertically. Therefore, there is little or no penetration across the shear interface (Figure 6.2). However, even in these cases, lateral roots can play an important role by maintaining the continuity of root-permeated soil mantle on a slope. For these cases, it is important to know or predict the extent of root spread. It is normally reported in relative multiples of tree height or crown radius. A useful rule of thumb is that a root system will spread out a distance at least equal to 1.5 times the radius of the crown, but this rule is strongly soil condition dependent. A high bulk density affects

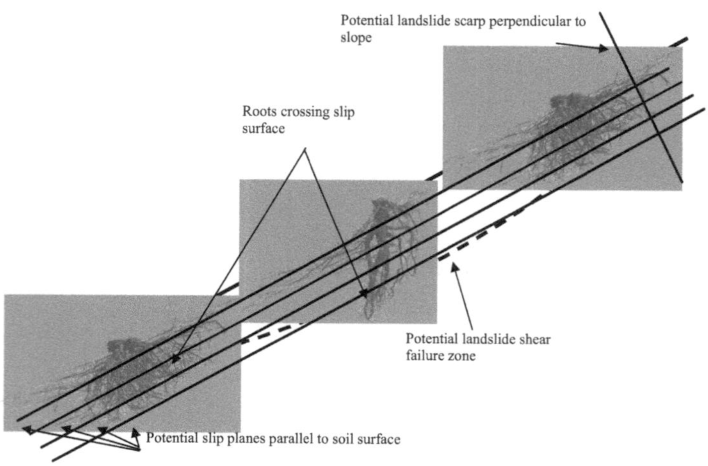

Figure 6-2. Roots crossing potential slip planes and shear surfaces will reinforce the slope against landslides (after Danjon et al. 2007, reprinted by permission of the publisher).

the root apex reducing the penetration across deeper soil layers. A general behavior to be noted is that root systems tend to grow wide and deep in well drained soils as opposed to developing a flat, plate-like structure in a surface soil underlain by a more dense (clay) or rocky substratum (Henderson et al. 1983).

With regard to protection forests, a large body of recent research exists on species suitability in sustaining rockfall damage. When trees are subjected to rockfall, they may uproot, suffer stem breakage, or kinetic energy may be transferred to the crown, causing the latter to break (see Chapter 7). Certain tree species, particularly angiosperms, appear to be more resistant to mechanical failure than others, often sustaining wounds only (Dorren et al. 2005; Stokes et al. 2005). Dorren et al. (2005) determined the efficiency of different alpine forest species to rockfall impacts. By calculating the energy dissipated during a rockfall impact for different species, Dorren et al. (2005) determined that the order in which species could dissipate the most energy, and hence were more resistant to rockfall was: *Quercus robur* > *Fagus sylvatica* > *Acer pseudoplatanus* > *Abies alba* > *Larix decidua*/*Picea abies*. In trees that do not fail but which are hit by falling rocks, wounds will be sustained which may lead to mortality. Mortality rates differ among tree species damaged by rockfall. It has been calculated that the mortality rate of Norway spruce (*P. abies*) increases by 66% after sustaining a rockfall wound, whereas in larch (*L. decidua*), the rate only increases by 23% (Brauner et al. 2005). The reason for this lower rate of mortality is because larch is a thicker-barked species than spruce. Thicker bark will help protect the internal part of the tree against low-energy rock impacts and can also grow faster around the new wound, thus accelerating the healing process. If wounds do not heal quickly, trees are more susceptible to attack by pathogens.

In protection forests, felled trees also serve a protective function. If felled and positioned correctly in rockfall corridors, logs, snags and windthrown trees can "catch" or deviate falling rocks into stands with a high stem density or channels with a high surface roughness such as depressions where rocks have accumulated (Kupferschmid Albisetti et al. 2003; Dorren et al. 2005; Schönenberger et al. 2005). When felled, the wood of certain species is more mechanically resistant and durable (resistant to pathogen decay over time) than others. By leaving felled snags and logs unharvested, it has been predicted in Norway spruce stands, that effective protection against rockfall activity and avalanche release will be provided for 30 years (Kupferschmid Albisetti et al. 2003). In experiments where the durability of felled logs in an Alpine forest was tested over several years, it was found that European beech (*F. sylvatica*) and silver birch (*Betula pendula* Roth.) were significantly less durable than Norway spruce or silver fir, (*A. abies*) with > 20% wood degradation in only two years (Stokes 2006). Therefore, by integrating such knowledge, a protection forest may consist of both living trees and felled stems to provide the ultimate protection against rockfall.

Table 6-4. Plants suitable for ecotechnological solutions in Europe (compiled from Schiechtl 1980; Coppin and Richards 1990; Lammeranner et al. 2005; Mattia et al. 2005; Mickovski et al. 2005; Stokes et al. 2005; Dorren and Berger 2006; Heumader 2007; Schmid et al. 2007 and see Plants for a future – species database (http://www.pfaf.org/index.html), 2000; USDA, NRCS).

Key to Function classification: AB – anchoring and buttressing of deep tap roots; BC – bank and channel reinforcement; DR – deep reinforcement and soil strength enhancement; PH – phreatophytes removing soil moisture, SP – surface protection, shallow reinforcement and erosion control. Key to Efficacy of rockfall: 0 – species is not suitable for protection from rockfall; + – limited rockfall protection; ++ – good rockfall protection; +++ – excellent rockfall protection.

A. Grasses.

Latin and English name	Region	Habitat/Properties	Altitude (asl)	Morphology: max. height (H) and max. root depth (R)	Pio-neer plant	Func-tion	Comments
Agropyrum cristatum L. (Crested wheatgrass)	Europe	Mountain slopes.	1200-2000 m			SP	Perennial. Suitable for highway vegetation. Can impede the spread of other grasses eg *Bromus tectorum*.
Agropyron repens (L.) Beauv. (Couch grass)	UK, Alpine	Arable weed, not to be used near arable land, needs light. Dry to moist, fertile soil.	up to 900 m	H 0.2-1.5 m R 0.8 m Rhizomatous		SP	Perennial.
Agrostis canina L. (Bent grass)	Alpine	Fens and bogs, peat areas, wet woodland, pioneer plant of open wet soil.	up to 1100 m	H 0.2-0.6 m R 0.2 m	Yes	SP	Perennial.
Agrostis capillaris L. (Common bent)	UK, Europe, West Asia	Dry heaths and grassland, preferring slightly acid soils.		H 0.7 m Rhizomatous		SP	Wide soil tolerance. Many cultivars available including those tolerant of heavy metal contamination. Perennial.
Agrostis castellana Boiss. & Reuter (Highland bent)	UK, SW & SE Europe	Grasslands.		Rhizomes		SP	Perennial.

Agrostis gigantea Roth (Black bent, red top)	Alpine	Riverbanks and lakeshore, wet woodland, unsuitable for lawns.	up to 1400 m	H 0.4-1.0 m R 0.3 m with short rhizomes		SP	Biennial-perennial.
Agrostis stolonifera L. (Creeping bent)	UK, Alpine	Riverbanks, ditches, wet areas, moisture indicator, dense sward, can stand heavy grazing.	up to 1800 m	H 0.1-0.7 m R 0.3 m Stoloniferous, rooting at the nodes	Yes	SP, BC	Spreads by stolons, wide soil tolerance. Many cultivars available but prefers damp soils; tolerates occasional flooding and salt. Perennial.
Agrostis tenuis L. (Common bent, brown top)	Alpine	Grassy patches in woodland, moist grassland in mountainous regions, moors, recently cut forest areas; indicates acid and very poor soil conditions; valuable grass for mountainous terrain.	up to 2200 m	H 0.2-0.4 m R 0.5 m		SP	Does well in grass mixtures. Perennial.
Alopecurus pratensis L. (Meadow foxtail)	Alpine	Riverbanks, alluvial area, resistant to late frosts, overwatering, tolerant to long lasting snow cover, needs fertiliser and irrigation, then suitable for poor soils; moisture indicator, loosely tufted.	up to 1800 m	H 0.3-1.0 m R 0.2 m		SP	Perennial, does not tolerate heavy grazing.
Ammophila arenaria (L.) Link (Marram grass)	UK	Sand dunes. Prefers light (sandy) and medium (loamy) soils, requires well-drained soil and can grow in nutritionally poor soil. Can grow in very acid soil. It cannot grow in the shade. Requires dry or moist soil. Drought tolerant. Salt tolerant.	Coastal	H 1.2 m Rhizomes with adventitous roots		SP	Perennial. Propagated by cuttings.

Anthoxantum odoratum L. (Sweet vernal grass)	Alpine	Meadows and pastures, open wood land, poor mountain meadows together with *Festuca rubra*, *Agrostis tenuis*; grows during winter, can stand irrigation, poor soil, short lived.	up to 2500 m	H 0.3-0.5 m R 0.5 m		SP	Perennial.
Arrhenatherum elatius (L.) Beauv. ex J.& K. Presl (Tall oat grass)	UK/Alpine	Moist and fertile meadows, at higher altitudes replaced by *Trisetum sp.*, main species of fertilised meadows, avoid wet heavy and compacted soils, sensitive to moist-cool conditions, loosely tufted.	up to 1500 m	H 0.5-1.5 m Roots strong, deep rooting		SP	Wide soil tolerance, natural coloniser of embankments and cuttings. Tall. Perennial.
Avena sativa L. (Oats)	N. Europe	Dry wasteland, cultivated ground and meadows, especially on heavier soils. Requires well-drained soil and can grow in heavy clay and nutritionally poor soils. Wide soil tolerance. It cannot grow in the shade. Drought tolerant.	up to 1600 m	H 0.8 m R 0.3 m		SP	Bi-annual to perennial.
Avenella flexuosa (L.) Drej. (Oat grass, flexible)	Alpine	Acid humus pioneer on poor soils, shade tolerant.	up to 2200 m	H 0.3 m, slender R 0.5m, strong	Yes	SP	Perennial.
Brachypodium pinnatum (L.) Beauv. (Heath/Chalk false brome)	Alpine	Indicates base rich soil and deteriorating woodland conditions, enhanced by burning, weakened by fertiliser application.	up to 1600 m	H 0.6-1.2 m R 0.5 m, strong		SP	Perennial.

Bromus erectus Huds. (Upright brome)	Alpine	Semi-arid grassland on limestone, in the south on gneiss and serpentine, intolerant of fertiliser and irrigation, avoids shade and wetness, but very resistant to dry heat, tufted.	up to 1400 m	H 0.6 m R 0.8 m, strong	Yes	SP	Perennial.
Bromus inermis Leyss. (Hungarian brome)	Alpine	Very drought and cold resistant.	up to 1200 m	H 0.3-1.4 m Rhizomatous, deep rooted	Yes	SP	Perennial.
Bromus mollis auct. non L. (Soft brome)	Alpine	Grassland weed, indicates poor soil, suitable cover crop on dry sites.	up to 1000 m	H 0.2-0.8 m R 0.2 m		SP	Annual.
Carex acutiformis Ehrh. (Lesser Pond sedge)	UK	Bogs. Prefers light (sandy), medium (loamy) and heavy (clay) soils. Wide soil tolerance. It can grow in semi-shade (light woodland) or no shade. It requires moist or wet soil.		H 0.75 m Rhizomes		BC	Perennial. Marginal zone (shallow water). Transplant fragments.
Carex riparia auct. non M.A. Curtis (Greater pond sedge)	UK	Marshes and bogs. Prefers light (sandy), medium (loamy) and heavy (clay) soils. Wide soil tolerance. It can grow in semi-shade (light woodland) or no shade. It requires moist or wet soil.		H 1.5 m Rhizomes		BC	Evergreen. Perennial. Marginal zone (shallow water). Transplant fragments.
Cynodon dactylon (L.) Pers. (Couch grass)	Alpine	Pioneer for the quick stabilisation of sandy soil in low altitudes, starts late in the season, frost hardy, pasture grass, turns brown in winter.	up to 1000 m	Stem procumbent with long stolons Rhizomatous	Yes	SP	Perennial.
Cynosurus cristatus L. (Crested dog's tail)	Alpine	Fertilised permanent pasture, indicates heavy soil, frost tender, shade tolerant for pastures and meadows.	up to 1500 m	H 0.3-0.6 m		SP	Perennial.

Dactylis glomerata L. (Cocksfoot)	Alpine	All round pioneer, vigorous when fertilised.	up to 1900 m	H 0.5-1 m Tufted R 0.4 m	Yes	SP	Perennial.
Deschampsia caespitosa (L.) Beauv. (Tufted hair grass)	Alpine	Intermittent moist locations in woodland and meadows, wet places and spring horizons, marshes, vigorous, stiffly tufted.	up to 2800 m	H 0.3-0.8 m R 1.0 m		SP	Perennial.
Festuca longifolia auct. non Thuill. (Hard fescue)	UK	Poor, well-drained shallow soils, wide soil tolerance. It can grow in semi-shade (light woodland) or no shade. Drought tolerant.	up to 200 m in UK and 1800 m in Pyrenees	H 0.25 m		SP	Perennial. Wear tolerant. Not rhizomatous.
Festuca arundinacea Schreb. (Tall fescue)	Alpine	Low fertility, acid, clay, loamy and sandy soils. Partial shade tolerant. Drought tolerant. Indicates soil compaction and wetness, tufted.		H 0.6-1.5 m Rhizomatous, deep rooted		SP	Perennial. Wear tolerant, suitable for pathways and terraces in vineyards and orchards.
Festuca ovina L. (Sheep's fescue)	Alpine	Dry poor grassland on soils derived from acid rocks, indicates degradation in forests.	up to 2300 m	H 0.15-0.4 m Dense, short R 0.5 m		SP	Perennial.
Festuca pratensis Huds. (Meadow fescue)	Alpine	Fertile meadows and pastures, prefers heavy moist soils, winter hardy, sensitive to over-use.	up to 1600 m	H 0.3-1.2 m Loose tufts Shallow rooted		SP	Perennial.
Festuca rubra L. ssp. *rubra* (Creeping red fescue)	UK/Alpine	Montane meadows and pastures, coniferous wood and deciduous forests, drought and wetness sensitive.	up to 2000 m	H 0.2-0.7 m R 0.5 m; Rhizomatous		SP, BC	Very wide soil tolerance. Many cultivars available including those tolerant of heavy metals and salt. Perennial.

Festuca rubra L. ssp. *commutata* Gaudin (Red fescue)	Alpine	On acid soils, replaced by *Nardus stricta* if over used.	up to 2000 m	H 0.1-0.6 m Dense sward R 0.5 m		SP	Perennial.
Festuca tenuifolia Sibthorp (Fine-leaved sheep's fescue)	Alpine	On acid soils, indicates deterioration in forests.	up to 1000 m	H 0.2-0.3 m Slender Shallow rooted		SP	Perennial.
Festuca trachyphylla (Hack.) Krajina (Rough leaved fescue)	UK and Central Europe	Sandy soils.	up to 1000 m	H 0.1-0.15 m Very shallow rooted		SP	Perennial.
Festuca valesiaca Schleich. (Volga fescue)	Alpine	Dry grasslands.		H 0.5 m Densely tufted		SP	Perennnial.
Glyceria maxima (Hartman) Holmb. (Reed sweet grass)	UK	Ponds, marshes. Wet soils.		H 0.1-0.25 m R 1 m Extensive, rhizomes		SP, BC	Spreading, can be invasive; tolerates damage. Marginal and emergent zones.
Holcus lanatus L. (Yorkshire fog)	UK, Alpine	Grassland, moorland. Indicates acid soils low in nitrogen, frost tender, green during winter, in years of good rainfall very prolific on poor soils.	up to 900 m	H 0.3-1.6 m Tufted R 0.4 m		SP	Perennial.
Holcus mollis L. (Creeping soft grass)	UK, Alpine	Arable lands and ploughed out pasture, wet soil areas, never on calcareous soil, sandy soils, troublesome weed in gardens and arable land, starts growing later than *Holcus lanatus*.	up to 1500 m	H 0.3-1.6 m R 0.4 m Long rhizomes		SP	Perennial.

Leymus arenarius (L.) Hochst. (Lyme grass)	UK	Sand dunes. Grows in poor soils. Drought tolerant.	Coastal	H 1.2 m		SP	Perennial. Salt tolerant.
Lolium multiflorum Lam. (Ryegrass)	UK, C. and S. Europe	Meadows, field margins and road verges; avoid acid soils.	up to 800 m	H 0.3 m	Yes	SP	Biennial.
Lolium multiflorum Lam. ssp. *italicum* Schinz & R. Keller (Italian ryegrass)	Alpine	Only in Atlantic areas under moist mild climatic conditions; frost damage below -5ºC, requires potash-rich soils; vigorous after cutting; unsuitable for permanent meadows.	up to 1700 m	H 0.3-0.9 m R 0.8 m		SP	Annual - bi-annual.
Lolium perenne L. (Perennial ryegrass)	UK/Alpine	Tolerant to repeated cutting and trampling; pioneer of well-aerated and moist soil, needs fertiliser; fast growing.	up to 1000 m	H 0.3-0.7 m Dense tufts R 1.2 m	Yes	SP	Many cultivars available. Perennial.
Lygeum spartum L. (Albardine)	Medi-terranean	Grows on rocky sea shores, dry sandy or clayey soils. Tolerates gypsum. Will grow on acid mine tailings.		H 0.45 m Thick, dense, fibrous root system extending from rhizomes		SP	Perennial.
Phalaris arundinacea L. (Reed canary grass)	UK	Wet and low fertility soils. Frost hardy.	Lowlands to mountain areas	H 2.0 m Rhizomes		DR, PH	No commercial seed, established as live plant and rhizome fragments.
Phleum pratense L. (Timothy)	Alpine	Pastures, tolerates cold climate, wetness and prolonged snow cover, wind; grazing increases vigour and yield.	up to 2600 m	H 0.2-1.0 m Loose tufts Roots delicate with short runners		SP	Perennial.
Poa annua L. (Annual meadow grass)	UK/Alpine	Tolerates trampling and heavy fertiliser applications.	up to 3000 m	H 0.02-0.35 m Dense low swards Shallow roots		SP	Annual - perennial.

Poa compressa L. (Flat-stalked meadow grass)	UK/Alpine	Not on soils derived from acid rocks, green over winter.	up to 1800 m	H 0.2-0.4 m Loose tufts with runners R 0.2 m		SP	Rhizomes tolerant of very infertile conditions. Perennial.
Poa nemoralis L. *(Wood meadow grass)*	Alpine	On heavy soils, grows early after melting snow, very shade tolerant, not to be planted in pure stands, does not form close sward.	up to 2300 m	H 0.3-1.2 m Tufted Shallow rooting		SP	Perennial.
Poa palustris L. (Swamp meadow grass)	Alpine	On riverbanks, early, resistant to late frost.	up to 1500 m	H 0.3-1.2 m Shallow rooting		SP	Perennial.
Poa pratensis L. (Smooth meadow grass)	UK/Alpine	Important constituent of meadows and pastures, wide ecological amplitude, hardy, long living, grows early in spring, very suitable for first seeding.	up to 2300 m	H 0.15-0.9 m Dense tufts R 0.65 m Rhizomatous		SP	Wide soil tolerance, wear tolerant. Many cultivars available. Perennial.
Poa trivialis L. (Rough meadow grass)	Alpine	On spring horizons, fens, sensitive to dry air and soil, frost and prolonged snow cover, resistant to heavy grazing, responds well to organic fertilisers, long life, forms dense sward.	up to 1600 m	H 0.3-0.9 m Surface runners, shallow rooted		SP	
Puccinellia distans (Jacq.) Parl. (sea meadow grass)	Alpine	Saline soils, manure heaps, cattle pens, solonetzic soil.	1000-2600 m	H 0.15-0.5 m Tufted R 0.25 m		SP	Perennial.
Trisetum flavescens (L.) Beauv. (Yellow oat)	Alpine	Fertile meadows in montane and subalpine regions, frost sensitive, resistant to repeated cutting.	up to 2300 m	H 0.3-0.8 m Loosely tufted R 0.4 m		SP	Perennial.
Secale cereale (L.) (Cereal rye)	UK/Europe	Is hardy and not frost tender. Prefers well-drained, light soil. Cannot grow in the shade. Can tolerate strong winds.		H 1.8 m		SP	Biennial. Rots to form mulch which protects soil surface.
Vetiveria zizanioides (L.) Nash. (Vetiver)	Tropics	Frost tender.		H 1.5 m Densely tufted R 3.0 m Branched and fibrous		DR	Perennial grass sterile outside its natural habitat.

B. Herbs and Legumes.

Latin and English name	Region	Habitat/Properties	Altitude (asl)	Morphology: max. height (H) and max. root depth (R)	Pioneer plant	Function	Comments
Alchemilla xanthochlora Rothm. (Lady's mantle)	UK/Europe	Woodlands and meadows. Prefers well-drained, neutral or basic soil.		H 0.3 m		SP	Perennial. Good ground cover. Immune to rabbit predation.
Achillea millefolium L. (Yarrow, milfoil)	UK/Alpine	Meadows and pastures, drought resistant, indicator of fertile soil. Salt tolerant.	up to 1900 m	H 0.2-0.6 m R 0.4 m Rhizomatous		DR	Perennial.
Anthyllis vulneraria L. (Kidney vetch)	Alpine	Frost and drought resistant.	up to 2000 m	H 0.1-0.5 m R >1.0 m	Yes	DR	Perennial, cannot take fertiliser application or irrigation.
Chrysanthemum leucanthemum L. (Ox-eye, dog daisy)	UK/Alpine	Pioneer on loose, well-aerated, immature soils, indicates poor fertility in meadows. Tall growth.	up to 2200 m	H 0.3-0.6 m R 0.6 m	Yes	DR	Perennial.
Coronilla varia L. (Crown vetch)	UK/Alpine	Dry, sunny slopes. Wide tolerance, dense growth, slow to establish particularly in the north.	up to 900 m	H 0.3-1.2 m R 0.9 m		SP; DR	Perennial.
Lotus corniculatus L. (Birds-foot trefoil)	Europe	Grasslands, meadows. Wide soil tolerance. Salt tolerant.	up to 2600 m	H 0.3-1.2 m R 0.6 m Taproot		SP; DR	Perennial.

Lotus uliginosus Schkuhr (Greater bird's foot trefoil)	Alpine	Semi-dry turf, fertile meadows and pastures, prefers calcareous soils, high temperature resistance.	up to 2300 m	H 0.05-0.6 m Taproot up to 1.0 m		SP; DR	Perennial, persists for 20 years.
Lupinus albus L. (White lupin)	S. Europe	Acid moist soil. Shade intolerant.	up to 600 m	H 0.2-1.0 m R 0.75 m		DR	Annual.
Lupinus luteus L. (Sweet lupin)	Italy and Medi-terranean	Acid soils. Moist soil. Shade intolerant.	up to 1400 m	H 0.3-1.2 m		DR	Annual.
Lupinus polyphyllus Lindl. (Garden lupin)	UK/ Alpine	On wood land fringes and clearings. Wide soil tolerance, will die out with several hard winters.	up to 1400 m	H 1.0-1.5 m R >1.0 m		SP; DR	Perennial.
Medicago falcata L. (Sickle medick)	Alpine	Not cultivated as it becomes woody.	up to 1100 m	H 0.2-1.0 m Deep rooted		DR	Perennial.
Medicago lupulina L. (Black medick)	Alpine	Dry meadows of good fertility, indicator of dry habitat, undemanding pioneer, prefers calcareous soil, frost resistant, can be heavily grazed.	up to 1500 m	H 0.1-0.6 m R 0.5 m Thin tap root	Yes	SP	Annual - perennial.
Medicago sativa L. (Lucerne)	UK/Alpine	Drought tolerant, neutral/alkaline soils. Sensitive to late frosts.	up to 1000 m	H 0.3 -1.2 m R 2.5-5.0 m (up to 10.0 m) Very rough taproot		SP; DR; PH	Perennial. Only available as hybrid.
Melilotus albus (L.) Lam. (White melilot)	Alpine	Drought resistant, becomes woody, needs mowing.	up to 1800 m	H 0.3-1.2 m R 0.7 m Thick taproot		DR	Bi-annual.

Melilotus officinalis (L.) Lam. (Common melilot)	Alpine	Grassed areas, roadsides.	up to 1000 m	H 0.3-1.0 m R 0.75 m Taproot		DR	Biennial. Medicinal.
Myosotis alpestris Schmidt (Alpine Forget-me-not)	UK, Europe, W. Asia.	Woodlands and meadows. Hardy to -20°. Prefers well-drained, moist gritty soils.		H 0.2 m		SP	Perennial. Not susceptible to grazing by deer.
Onobrychis viciifolia Scop. (Common sainfoin)	UK/Alpine	Dry soil indicator in *Arrhenatheretum*, important feed plant on dry, clayey calcareous soil, sensitive to grazing. Neutral/alkaline soils, drought tolerant.	up to 2000 m	H 0.1-0.7 m R 1.0-4.0 m		SP; DR	Lasts 4-6 years.
Phacelia tanacetifolia Benth.	Europe	Open flats and slopes. Moist soil. Shade intolerant.	up to 1000 m	H 0.7 m R 0.2 m		SP	Useful as green crop, annual.
Pimpinella saxifraga L. (Burnet saxifrage)	UK/Alpine	Poor, dry soil.	up to 2300 m	H 0.15-0.5 m R 1.3 m, Roots 8-10 m long		DR	Perennial.
Pisum sativum L. (Garden pea)	Europe	Waste ground. Well drained soil.	up to 1000 m	H 0.5-2.0 m Tapering taproot		DR	Useful as green crop, annual.
Plantago lanceolata L. (Ribwort plantain)	Worldwide	Many soil types throughout the world. Short growth.	up to 1800 m	H 0.05-0.5 m R 0.6 m	Yes	DR	Perennial.
Sanguisorba minor Scop. (Salad burnet)	UK/Alpine	Tall growth; prefers lime soils but tolerates infertile soils. Grasslands.	up to 1200 m	H 0.3-0.6 m R 1.5 m	Yes	DR	Perennial.

Trifolium dubium Sibthorp (Lesser clover)	UK/Alpine	Fertile meadows and pastures, needs heavy nitrogen dressings.	up to 1000 m	H 0.05-0.35 m R 0.2 m		SP	Annual-bi-annual.
Trifolium hybridum L. (Alsike clover)	UK/Alpine	Pioneer on moraines, tolerates moist and cool conditions, frost resistant, tolerates prolonged snow cover, sensitive to shade and dry soil conditions. Tolerates waterlogging.	up to 2000 m	H 0.2-0.7 m R 0.2-0.8 m Branched root system	Yes	SP; DR	Bi-annual - perennial.
Trifolium pratense L. (Red clover)	UK/Alpine	Fertile meadows and pastures, but also moist and poor meadows, sensitive in spring to grazing pressure, important feed plant.	up to 2200 m	H 0.2-1.2 m R 2.0 m Branched		SP; DR	Perennial.
Trifolium repens L. (White clover)	UK/Alpine	Heavily used turf, meadows, parks, aerodromes in humid areas, very prolific. Requires moderate fertility.	up to 2300 m	H 0.1-0.5 m R 0.7 m Rooting at nodes		SP	Perennial. Many cultivars available.
Veratrum album L. (White hellebore)	Europe/ Alpine	Prefers deep, fertile, moist soil in woodlands or meadows.		H 1.5 m		SP	Perennial. Highly poisonous and not susceptible to grazing by deer.
Vicia sativa L. (Common vetch)	UK/Alpine	Valuable cover crop.	up to 1600 m	H 0.3-1.0 m Prone or climbing R 0.5 m		SP	Annual; many varieties.
Vicia villosa Roth (Fodder vetch)	UK/Alpine	Cover crop. Frost hardy if sown early.	up to 1700 m	H 0.3-0.6 m Prone or climbing R 0.6 m		SP; DR	Annual - biennial.

C. Shrubs and small trees.

Latin and English name	Region	Habitat/Properties	Altitude (asl)	Morphology: max. height (H) and max. root depth (R)	Pioneer plant	Efficacy against rockfall	Function	Comments
Acer campestre L. (Field maple)	Northern Europe	Broadleaved forests and spinneys. Tolerates only light shade.	up to 800 m	Slow growing shrub to small tree. Strong heart-shaped root system.		+	SP	
Alnus incana (L.) Moench (Grey Alder)	Europe, submontane to montane	Woodlands. Heavy clay soil. N-fixer.	500-1600 m	H 18.0 m	Yes	+	PH	Rapid coloniser. Used for vegetated crib walls and hedge brush layers.
Alnus viridis (Vill.) Lam. & DC. (Green alder)	Alpine	Forms closed formations on montane and subalpine alluvial flood plains and shelves. Wet, lower slopes in snow rich areas.	500-2000 m	H 7.0 m Shallow rooted	Yes	+	SP	Used for vegetated crib walls and hedge brush layers.
Amelanchier ovalis Medik. (Snowy mespilus)	Central and Southern Europe, alpine	Occasionally in sunny oak and pinewoods; on rocky, stony slopes. Likes limestone. Deciduous showy shrub.	up to 1800 m	H 5.0 m Roots in rock fissures, widespread root system		0	SP	
Atriplex halmus (Sea orach)	Mediterranean	Coastal sands and saltmarshes. Evergreen shrub. Resistant to drought, extreme temperature and salinity.		H 2.0 m Taproot		0	SP	

Berberis vulgaris L. (Common barberry)	Submontane	Light deciduous woodland, hedges, roadsides, clearings etc, preferring a sunny position and a chalky soil.	up to 1800 m	H 3.0 m Suckers from roots		0	SP	Used for vegetated crib walls.
Clematis vitalba L. (Traveller's joy)	Submontane	Hedgerows and wood margins, usually on calcareous soils.	up to 1000 m	Climber. Roots like shade. Roots from stems touching the ground.		0	SP	Vigorous, destructive plant, kills native species.
Colutea arborscens L. (Bladder senna)	S. Europe – Mediterranean. Naturalised in Britain.	In sunny, warm locations of oakwoods in the south, open woods, roadsides, railway banks, on any soil type. Deciduous shrub. Tolerates strong winds but not maritime exposure. Tolerates atmospheric pollution. N-fixer.	up to 800 m	H 3.6 m Deep rooted		0	SP; DR	
Cornus mas L. (Cornelian cherry)	Alpine	Sunny, open woodland and woodland fringes. Tolerates light shade.	up to 600 m	H 4.0 m shrub or 6.0 m tree Widespread strong root system		0	SP	
Cornus sanguinea L. (Dogwood)	Alpine	Sunny, open broadleaved mixed woodland, wood verges. Any soil type, light shade only.	up to 1000 m	H 3.5 m Spreads suckers. Widespread strong root system.		0	SP	Used for vegetated crib walls.
Corylus avellana L. (Hazel)	Sub-mediterranean/ sub-atlantic	Secondary formations in the temperate broadleaved mixed woodland zone in sunny, sub-mediterranean/sub-atlantic locations. Many soil types, tolerates moderate shade.	up to 1400 m	H 5.0-6.5 m Shrub or small tree, spreading habit Extensive strong root system		++	SP	

Cotoneaster microphyllus Lindl.	UK (E. Asia – Himalayas)	Trailing on rocks or spreading on grassy hillsides. Wide soil tolerance. Evergreen. Tolerates strong winds but not maritme exposure. Tolerates atmospheric pollution.	1200-5400 m	H 1.0 m by 2.0 m at a slow rate		0	SP	
Crataegus monogyna Jacq. (Common hawthorn)	UK, Sub-mediterranean/ sub-atlantic	Broadleaved and coniferous woodlands. Can be coppiced; wide soil tolerance. Salt tolerant.	up to 1000 m	H 6.0-10.0 m Thorny Deep rooted (not in clay soils)		+	DR; PH	Can reach 100 years of age.
Cytisus scoparius (L.) Link (Broom)	UK	On acid, non-calcareous soil on sunny slopes in areas of mild winters. N-fixer, salt tolerant.	low lying areas to 1100 m	R 2.0 m Extensive root system	Yes	0/+	DR	
Erythrina arborescens Roxb.	Nepal, India, Burma	Mountain valleys, grassy slopes. Deciduous tree. Can be raised from cuttings.	450-2100 m			0/+	SP	Used in palisades.
Euonymus europea L. (Spindle tree)	Alpine	Thickets and woodland fringes, open woodland. On calcareous soil and sunny positions, indicates loamy soil, semi-shade.	low lying areas to 1100 m	H 4.0 m Shrub or small tree Extensive root system		0/+	SP	
Frangula alnus P. Mill. (Alder blackthorn)	Alpine	Bogs and damp woods, riverbanks, alluvial areas, open oak and pinewoods. Tolerates flooding and oxygen poor soil due to compaction.	up to 1000 m	H 2.0-7.0 m Loosely erect, occasionally small tree Shallow rooted with root bulbils		0	SP	
Genista germanica L. (German greenweed)	C. and W. Europe. Peri-alpine	Acid soil of sunny slopes in the peri-alpine area. Thickets, poor pastures and heaths. Tolerates drought.	up to 750 m	H 1.5 m Spreading root system		0	SP	

Genista tinctoria L. (Dyer's greenweed)	UK/Alpine	Poor soils on sunny slopes in peri-alpine areas, infertile meadows, oak forests. Prefers heavy soil. N-fixer.	lowlands to 750 m	H 1.0 m R 1.0 m		0	DR	
Hedera helix L. (Ivy)	UK	Good groundcover as well as climber.				0	SP	
Hippophae rhamnoides L. (Sea buckthorn)	UK/Alpine	Sandy soils especially dunes. Tolerant of wind and salt. N-fixer.	up to 1000 m	H 3.0 m either small trees or thicket Rhizomes	Yes	0	SP; DR	
Laburnum alpinum (Mill.) Bercht. & Presl. (Alpine laburnum)	C. and S. Europe. Submontane to subalpine	Montane and subalpine beech-fir tree woodland in humid areas, in the Alps replaces green alder, vegetatively propagated. Tolerates strong winds but not maritme exposure. Tolerates atmospheric pollution.	500-1900 m	H 5.0 m by 6.0 m at a fast rate, ascending branches	Yes	0/+	SP	
Laburnum anagyoroides Med. (Golden chain)	Sub-mediterranean/ submontane/ subalpine	Oak woods and pine forests. Fertile and calcareous, humus, sandy to stony loamy soils in not too dry locations, mild winters. Soil improver, can be vegetatively propagated.	up to 2000 m	H 8.0 m Large shrub, ascending branches Strong widespread root system	Yes	0/+	SP	
Ligustrum vulgare L. (Privet)	Alpine	Sunny broadleaved and coniferous woodland on neutral- to base-rich soil, vegetatively propagated, can be clipped.	up to 1000 m	H 3.0 m Extensive roots, runners	Yes	0/+	SP	
Lonicera xylosteum L. (Fly honeysuckle)	Alpine	Broadleaved, coniferous mixed woodland. Tolerates moderate salinity and shade.	up to 1600 m	H 2.0 m Loosely spreading Shallow rooted	Yes	0/+	SP	

Lupinus arboreus Sims (Tree lupin)	UK	Woodland, coastal dunes. N-fixer, intolerant of exposure or hard winters. Salt tolerant.		H 1.5 m		0	SP; DR	Life span 7 years.
Lupinus perennis L. (Perennial lupin)	UK	Dry sandy soils. N-fixer.		H 0.6 m		0	SP; DR	
Pinus mugo Turra (Dwarf mountain pine)	Europe. Montane to subalpine	Woodlands. Well drained soils.	up to 2300 m	H 4.5 m	Yes	+/++	SP	
Pistacia lentiscus L. (Mastic tree)	Europe – Mediterranean	Open woods and scrub on dry hillsides. Evergreen shrub.		H 6.0-8.0 m Tap root		+/++	SP	
Prunus mahaleb L. (St. Lucia cherry)	Alpine	Sunny slopes and oak-pine forest in warmer areas. Semi-shade.	up to 800 m	H 4.0 m Loosely spreading, rarely a small tree Deep rooted		+	DR	
Prunus spinosa L. (Blackthorn)	UK/Alpine	Sunny positions, woodland fringes and open woodlands. Spiny. Forms dense thickets. Salt tolerant.	up to 1000 m	H 2.0-3.0 m Intricately branched Root suckers	Yes	0/+	DR	
Rhamnus cathartica L. (Buckthorn)	Alpine	Singly in wood verges and sunny spinneys. Mostly on calcareous soil.	lowlands up to 1400 m	H 2.0-3.0 m Much branched, thorny shrub or 6.0 m small tree, slow growing Extensive root system		0/+	SP	Used for vegetated crib walls.

Ribes alpinum L. (Mountain currant)	Submontane to subalpine	Hedges, woodlands, cliffs. Requires well-drained moist soil.	up to 1900 m	H 1.2 m		0	SP	
Ribes petraeum Wulf. (Flowering currant)	West and Central Europe	Woodlands. Wide soil tolerance.	up to 1900 m	H 1.8 m		0	SP	
Rosa canina L. (Dog rose)	Alpine	Sunny positions, spinneys, on many soil types.	up to 1350 m	H 3.0 m Loosely spreading. Deep rooted		0	SP	
Rosa rubiginosa L. (Sweet-briar)	UK/Alpine	Cosmopolitan in sunny positions, spinneys and woodland in warmer areas, prefers calcareous soils, indicator of loamy soil.	up to 1350 m	H 3.0 m Loosely spreading, thorny. Deep rooted		0	DR	
Rosa rugosa Thunb. (Shrub rose)	UK	Foothills to montane. Spiny. Salt tolerant.	up to 1000 m	Thorny		0	DR	
Rubus fruticosus L. (Blackberry)	UK/Alpine	Forest floor pioneer in humid areas with mild winters.	lowlands up to 1600 m	H 1.0 m Arching branches, fast spreading. Root shoots	Yes	0	SP	
Salix appendiculata Vill. (Goat willow)	UK, C. Europe-Balkans	From peri-alpine areas to the tree line in humid location on base-rich, neutral to slightly acid soil or calcareous rubble. Vegetatively propagated.	500-2100 m	H 6.0 m	Yes	+	SP; DR; PH	Coloniser of difficult sites including landslides on calcareous formations.
Salix aurita L. (Round-eared willow)	Alpine	From lowland to the montane regions, rare in areas of pronounced continental climate, marshes, on acid peaty gley soils.	up to 1600 m	H 2.5 m	Yes	0/+	SP; BC	

Salix cinerea L. (Grey sallow)	UK/Alpine	From coastal marshes to mountainous regions, in the warmer areas in drying marshes, fens. Alder coppice, on fertile acid sandy soils to clay soils. Tolerates water-logged conditions (gley soils).	up to 800 m	H 2.0-3.0 m	Yes	0/+	SP; DR; PH	Not good for channel work. Resistant to flooding.
Salix elaeagnos Scop. (Hoary willow)	Submontane to montane (subalpine)	Alluvial shelves of alpine river valleys; calcareous rubble just above the ground water table, but periodically dry, sandy-stony slides on steep slopes.	up to 1850 m	H 6.0 m rarely 15.0 m	Yes	+	SP	
Salix foetida Schleich. Ex. DC.	Alpine - subalpine	Woods, riverbanks and wet slopes, on acid, fertile, often marshy soil, by preference on moraines of siliceous rocks.	1700-2800 m	H 1.5 m		0/+	SP; BC	
Salix glabra Scop.	Montane to subalpine	Limestone mountains of the eastern Alps, on rubble, stony slopes and gullies, only on calcareous and dolomitic rocks.	1400-2000 m	H 1.5 m	Yes	0/+	SP	
Salix glaucosericea Flod.	Subalpine - Alpine	Scattered in the sub-alpine region of the central Alps in green alder woods and scrub on wet siliceous moraines and alluvial shelves.	1700-2000 m	H 1.5 m		0/+	SP	
Salix hastata L. (Mountain willow)	Montane and Alpine	In the high montane and subalpine regions of the Alps in willow-green alder scrub-bush areas in moist and shaded locations, neutral to weakly acid, fertile soils on various parent rocks. Tolerates prolonged snow cover.	1600-2400 m	H 3.0 m Rigid branches	Yes	+	SP	

Salix hegetschweileri Heer.	Subalpine and montane	Wet, usually weakly calcareous but fertile soils subject to percolating water, riverbanks and wet lower slopes of bayleaf willow bush areas.	1600-2000 m	H 4.5 m	Yes	+	SP	
Salix helvetica Vill. (Swiss willow)	Subalpine	Shaded scree slopes with dwarf shrubs and green alder scrub, non calcareous, wet rubble, skeletal soils. Tolerates prolonged and deep snow cover.	1700-2600 m	H 1.5 m		0/+	SP	
Salix mielichhoferi Saut.	Alpine	Wet slopes and riverbanks in the montane and subalpine zone on fertile skeletal soils.	1300-2200 m	H 4.0 m		+	SP	
Salix nigricans auct., non Sm. (Dark-leaved willow)	Subalpine	Moist, neutral to weakly acid clayey, gravelly or sandy soil, particularly in the cool-humid limestone areas.	up to 2400 m	H 8.0 m bush or small tree Adventitious roots in response to flooding	Yes	+	SP; BC	
Salix nigricans ssp. *alpicola* (Buser) D. & E. Lautenschlager	Alpine	The sub-species *alpicola* forms closed stands in the high montane and subalpine regions of the central Alps. Tolerates waterlogged conditions.				0/+	SP	
Salix purpurea L. (Purple osier)	Alpine	Softwood, alluvial bush, often in the pioneer stage, periodically flooded, usually calcareous alluvial, silty, sandy, gravelly soils. Slow growing.	up to 2300 m	H 6.0 m Extensive root system	Yes	+	DR; PH; BC	Most suitable willow for vegetative protection measures.

Salix repens L. (Creeping willow)	UK. Submontane	Wet heaths and moorlands. Moist soil. Deciduous shrub. Ground cover.	up to 1000 m	H 1.5 m Adventitious roots in response to flooding	Yes	0	BC	
Salix triandra L. (Almond-leaved willow)	UK/Peri-alpine	Softwood alluvial woodland, edges of rivers and ponds, periodically flooded, wet, often calcareous, silty, sandy or gravelly soils. Not very tolerant to shade.	lowlands to 1500 m	H 9.0 m at a fast rate. R 2-4 m	Yes	+	BC; PH; SP	Good for river works, bank repairs. Keep coppiced to prevent larger stems which obstruct flow.
Salix viminalis L. (Common osier)	UK/Peri-alpine	River valleys. Intermittently wet, base-rich and fertile silty-loamy sands. Grows as thicket.	up to 1400 m	H 5.0 m	Yes	+	BC; PH	Good for river works, bank repairs. Keep coppiced to prevent larger stems which obstruct flow.
Salix waldsteiniana Willd.	Subalpine	Moist, neutral to weakly acid, base-rich, loamy, skeletal soils. Tolerates prolonged and deep snow cover.	1400-2200 m	H 1.5 m		0/+	SP	
Sambucus nigra L. (Elder)	UK/Alpine	Damp woodland, wasteland, spinneys. Damp and fertile soil. Indicates nitrogen-rich soil.	up to 1500 m	H 5.0 m Much branched, wide crown, shrub or small tree Shallow rooted		+	SP	

Sambucus racemosa L. (Alpine elder)	Alpine	Damp and shady woods, cleared woodland, spinneys, in the montane region. Non-calcareous soils, nitrate indicator.	lowlands to 1800 m	H 3.0 m Tall spreading bush Shallow rooted, new shoots produced from roots		0/+	SP	
Tamarix L. (Tamarisk)	UK	Hedges and riverbanks. Wind and salt tolerant.		H 4.5 m Deep rooted		+	SP; DR; PH	
Ulex europaeus L. (Gorse)	UK	Moors, commons and heaths. Dry soils. N-fixer.		H 1.5 m	Yes	0	DR; BC	Soil stabilisation of sandy substrates. Fire risk if growing in stands on its own.
Viburnum lantana L. (Wayfaring tree)	Alpine - subalpine	Scattered in open pine and oak woods, spinneys on calcareous soils. Tolerates severe cutting.	up to 1400 m	H 4.0 m Extensive spreading root system		0/+	SP	Used for vegetated crib walls.
Viburnum opulus L. (Guelder rose)	Alpine	Alluvial shelves, bushes and broadleaved woods; hydromorphic deep alluvial soil. Cut branches shoot readily, indicates moving water table.	up to 1000 m	H 5.0 m Fast growing, large bush or small tree Extensive shallow root system		0/+	SP	

D. Propagated trees.

Latin and English name	Region	Habitat/Properties	Altitude (asl)	Morphology: max. height (H) and max. root depth (R)	Pioneer plant	Efficacy against rockfall	Function	Comments
Populus nigra L. (Black poplar)	Southern Alpine region	Softwood alluvials, hydromorphic but well aerated sandy silty soils, periodically flooded.	up to 1400 m	H 30.0 m Plate-type root system	Yes	+	SP	
Salix alba L. (White willow)	UK, subalpine-alpine	Softwood alluvials, lowland/lower montane regions, neutral fertile and calcareous alluvial sandy loams and loamy sands subject to periodic flooding, tolerates silt aggradation. Salt tolerant.	up to 1300 m	H 20.0 m Roots at water level	Yes	+	PH	
Salix alba L. ssp. *vitellina* (L.) Arcang. (Golden willow)	Alpine	Cultivated willow with yellow or reddish-orange young twigs. Ornamental.		H 20.0 m		+	BC	
Salix caprea L. (Sallow)	UK, Subalpine	Near lakes and streams. Wet soil.	up to 1700 m	H 10.0 m	Yes	+	SP; DR; PH	
Salix daphnoides (Violet willow)	Submontane to subalpine	Softwood alluvials of mountain streams, particularly in the montane zone of the limestone Alps; on loamy, gravelly-sandy, neutral- to base-rich alluvial soils.	up to 1850 m	H 15.0 m	Yes	+	SP; DR	
Salix fragilis L. (Crack willow)	Alpine - subalpine	Permanently wet soils, poor base status, moving water table in areas with cool summers; alluvials in higher lying areas, can tolerate stagnant water and gley conditions.	up to 1100 m	H 10.0-25.0 m	Yes	+	SP; DR	
Salix pentandra L. (Bay-leaved willow)	Alpine - subalpine	Coppices and alluvial woodland on wet, slightly acid alluvial soil of restricted permeability, mainly in lowlands and inner-alpine valleys.	up to 1800 m	H 12.0 m	Yes	+	SP; DR	
Salix rubens Schrank (Hybrid crack willow)	Alpine	Softwood alluvials.		H 25.0 m		+	SP; DR; BC	

E. Conifers.

Latin and English name	Region	Habitat/Properties	Altitude (asl)	Morphology: max. height (H) and max. root depth (R)	Pioneer plant	Efficacy against rockfall	Function	Comments
Cupressus macrocarpa Hartw. ex Gord. (Monterey cypress)	UK	Evergreen. Salt tolerant. Drought tolerant. Wide soil tolerance.		H 25.0 m		++	DR; PH	Fast growing.
Larix decidua P. Mill. (European larch)	UK/Alpine	Coniferous forests, also mixed woodland, pure stands in certain localities in the Western and Southern Alps. Main distribution in the alpine-continental Spruce and Arolla pine forests.	up to tree line at 2100-2400 m	H 35.0 m Deciduous taproot system	Yes	++	DR	
Picea abies (L.) Karst. (Norway Spruce)	Montane and subalpine	Pure and mixed stands on moist, humic, slightly acid soil derived from siliceous and calcareous rocks. Evergreen.	800-1900 m	H 35.0 m Dominant, wide-spread shallow roots, in very deep soil some vertical roots		++	SP; AB in certain deep soils	
Pinus nigra Arnold (Corsican and Austrian pine)	Europe	Mountains. Evergreen. Calcareous soils.		H 30.0 m Taproot system		++	DR; PH	

Pinus sylvestris L. (Scots pine)	UK/Alpine - subalpine	Mixed and multi-species pine forests on base-rich calcareous soil and in acid pine forest on poor acid soil. Drought resistant, frost hardy. Evergreen.	up to 1900 m	H 20.0 m Taproot system	Yes	++	DR; AB	
Pinus uncinata Mill. ex Mirb. (Mountain pine)	Submontane to subalpine	Montane and subalpine forest on very shallow, stony calcareous soil. Optimum in Western Alps, Pyrenees. Locally in small patches in montane and subalpine boggy areas. Evergreen.	up to tree line at 2000-2400 m	H 20.0 m Taproot system	Yes	++	DR; AB	

F. Broadleaf trees.

Latin and English name	Region	Habitat/Properties	Altitude (asl)	Morphology: max. height (H) and max. root depth (R)	Pioneer plant	Efficacy against rockfall	Function	Comments
Acer platanoides L. (Norway maple)	Europe	Mixed woodland. Moderately acid and immature soil. Tolerates atmospheric pollution.	Lowland to 1100 m	H 30.0 m Deep rooted, heart shaped root system		+++	DR	Inhibits growth of other plants.
Acer pseudoplatanus L. (Sycamore)	Europe	Moist, cool, mixed broadleaved tree woodland and hedgerows, requires humus-rich soil adequately supplied with water. Tolerant to long periods of gravel aggradation and exposure.	peri-alpine hills to 1700 m	H 25.0 m Deep rooted, heart shaped root system		+++	DR; AB	Used for vegetated crib walls.

Alnus glutinosa (L.) Gaertn. (Common alder)	Alpine	Forest pioneer of fens and riverbanks, derelict land. Acid, moist to wet soil, immature soil; N- fixer. Can be coppiced.	0-1500 m	H 20.0 m Root development depending on habitat, but always deeper than grey alder; adventitious roots in response to flooding	Yes	+	PH; DR; BC	Resistant to flooding.
Alnus incana (L.) Moench (Grey alder)	Alpine - subalpine	Alluvial immature soils, riverbanks in montane region. Forms single species woodland. N- fixer. Can be coppiced, drier sites.	0-1600 m	H 20.0 m Shallow rooted, heart shaped root system	Yes	+	SP; PH	Suitable for providing early rapid growth.
Betula pendula Roth (Silver birch)	Central Europe	Sandy and poor soil, sub-dominant in open spruce-oak-beech-alder woodland. Tolerant of infertile conditions.	up to 1800 m	H 20.0 m Intensive shallow root system	Yes	+	SP; DR	Used for vegetated crib walls.
Betula pubescens Ehrh. (Hairy birch)	Montane to subalpine	Acid, wet peaty soil and immature soils on siliceous parent material in humid areas.	up to 2100 m	H 12.0 m Intensive shallow root system	Yes	+	SP	
Carpinus betulus L. (Hornbeam)	Temperate Europe	Prefers low lying rich soils. Shade tolerant.	up to 1000 m	H 30.0 m Heart shaped root system		++	AB	
Castanea sativa P. Mill. (Sweet chestnut)	Europe	Woodlands. Well drained soil.	up to 1000 m	H 30.0 m Heart shaped root system		++	SP	
Fagus sylvatica L. (Beech)	Europe	Woodlands. Wide range of soil types, prefers chalky soil. Tolerates strong winds and atmospheric pollution.		H 40.0 m Heart shaped root system		+++		

Fraxinus excelsior L. (Ash)	UK/ Alpine	May form pure stands in hardwood forests and in gorges. Sensitive to late frost, important soil stabilising properties.	up to 1400 m	H 35.0 m Extensive, deep root systems, strong roots	Yes	++	DR	
Fraxinus ornus L. (Flowering ash)	Alpine	Warmer zones of broad leaved woodlands in the Rhine valley, lower slopes of the southern Alps and Pannonian areas.	up to 800 m	H 8.0 m Deep rooted		++	DR	
Populus L.	Europe	Open woodland. Wide soil tolerance.	up to 1400 m	H 20.0 m Heart shaped root system	Yes	+	BC	Resistant to flooding.
Populus alba L. (White poplar, abele)	Sub-montane	Woodland. Salt tolerant.	up to 800 m	H 30.0 m Extensive deep root system	Yes	+	PH; DR	
Populus tremula L. (Aspen)	Europe, Montane	Open woodland, heathland. Wide soil tolerance. Tolerates strong winds.	up to 1400 m	H 20.0 m Extensive and aggressive root systems	Yes	+	DR	Damage to buildings.
Prunus avium (L.) L. (Wild cherry, gean)	Alpine	Open broadleaved woodland. At higher altitudes only on forest margins.	up to 1700 m	H 15.0 m Deep rooted, heart shaped root system		+	DR	
Prunus padus L. (Bird cherry)	Sub-montane to subalpine	Mixed broadleaved woodland at higher altitudes. Intolerant to shading. Rich and fertile soils, resistant to flooding and gravel aggradations.	up to 1700 m	H 15.0 m Extensive root system with strong roots		+	DR	
Quercus petraea (Mattuschka) Liebl. (Durmast, sessile oak)	Sub-montane	Woodland. Acid soils. Tolerates strong winds.	up to 1000 m	H 40.0 m Heart shaped root system		+++	DR	

Quercus robur L. (Common oak)	Europe	Salt tolerant.	up to 1200 m	Deep taproot, heart shaped root system	Yes	+++	DR; AB; PH	
Sorbus aria (L.) Crantz. (Common whitebeam)	UK, Alpine	Sunny positions on chalk or limestone soils in broadleaved and coniferous woodland in warmer areas. Tolerates maritime exposure and atmospheric pollution.	lowland to 1500 m	H 12.0 m by 8.0 m Deep rooted	Yes	+	DR	Reaches age of 200 years.
Sorbus aucuparia L. (Rowan)	UK, Alpine	Wide-spread in almost all humid forest types. Tolerates light shade only, on any soil type.	lowland to 2000 m	H 15.0 m Deep rooted on deep soil, otherwise shallow		+	DR	Used for vegetated crib walls.
Tilia cordata P. Mill. (Small leaved lime)	Foothills to montane	Woodlands. Fertile limestone soils.	up to 1450 m	H 30.0 m Suckers, heart shaped root system		++	SP; DR	
Ulmus glabra Huds (Wych elm)	Foothills to montane	Woods, hedgerows, streams. Tolerates maritime exposure and atmospheric pollution.	up to 1400 m	H 30.0 m Heart shaped root system		++	SP	
Ulmus minor Salisb. (English elm)	Foothills to sub-montane	Hedgerows, road verges. Tolerates maritime exposure and atmospheric pollution.	up to 600 m	H 35.0 m	Yes	++	SP	

1.2 Plant selection for erosion control

The mechanics of how plant roots reinforce the soil are twofold. First, roots and root remnants physically bind soil particles and in this way form mechanical barriers for soil and water movement (Tengbeh 1993). Major controlling parameters of the mechanical influence of roots are: diameter, density, degree of bifurcation, appearance of root hairs, friction between root and soil and, obviously, root network distribution (Abe and Ziemer 1991; Gyssels et al. 2005; De Baets et al. 2007; Reubens et al. 2007). Shallow interlocking root networks can substantially contribute to mechanical reinforcement of soils, acting as an anchored net of densely interwoven roots (Sidle et al. 1985; Preston and Crozier 1999). Dense root mats carpet the ground and provide substantial soil cohesion, which ultimately limits erosion by overland flow (Prosser et al. 1995). Moreover, living and dead root systems can provide subsurface water flow pathways by creating biopores and thus reducing the amount of erosive overland flow. Secondly, roots and root remnants excrete binding agents and form a food source for micro-organisms that in turn produce other organic bindings (Reid and Goss 1987). These bindings increase the amount of stable soil aggregates in the long term and thus reduce soil erodibility (Hartman and De Boodt 1974).

Nevertheless, it can be assumed that a shift in erosion control occurs within the growing cycle of plants, because of changes in plant characteristics. In the early plant stages, plant shoots are limited in number and they are very flexible, whereas emerging plant roots can contribute to soil cohesion, can provide additional strength and can form a physical barrier for flowing water. Over time, shoots will progressively become more dominant in reducing soil erosion as the number, height, continuity and stiffness of the plant shoots increases. These findings stress the temporal character of the relative influence of different parts of the vegetation on soil erosion rates by concentrated flow, as controlled by the growing cycle of plants. It is obvious that the influencing role of plant roots on concentrated flow erosion will largely depend on root type and their spatial distribution, as suggested by a study of Dissmeyer and Foster (1985). These authors show that erosion rates decline exponentially with an increase in surface soil occupied by fine roots, and that this effect is more pronounced in the case of fibrous lateral roots. These, in contrast with tap roots, form an important network just below the soil surface that reinforces the strength of the soil.

Cereal and grass roots are of the fibrous root type with fine diameters (ca. 0.24 and 0.15 mm) (Van Noordwijk and Brouwer 1991) and produce a dense root mat just below the soil surface. Winter wheat (*Triticum aestivum* L.) has one of the most prolific root systems of all arable crops (Barraclough et al. 1991) and could consequently be capable of controlling erosion in arable fields that are prone to soil erosion by concentrated flow if sown at sufficiently high densities. It is obvious that the date of the first intense rains on the recently sown seedbeds is crucial with respect to possible impacts of the roots on soil erosion rates.

In order to develop a certain rooting volume, roots need some time. Moreover, the degree of root distribution in the soil is influenced by the date of sowing: winter cereals produce more dry root matter than spring crops due to their longer growing season and this also applies to their shoot systems (Barraclough et al. 1991). For this reason, the technique of using root density as an erosion control strategy (e.g. Gyssels et al. 2002, 2005) will be especially useful in winter periods, when many fields are left bare after the last field operation in autumn. Plant roots are crucial in the early plant stages, when the aboveground biomass is fairly limited. If at this particular erosion-sensitive time plant roots are well developed, their rooting network can possibly temper soil erosion by water. The degree of soil erosion reduction by roots will be strongly conditioned by their spatial arrangement and rooting characteristics. In order to obtain a good rooting biomass, one could possibly sow at a higher than usual rate in zones at risk of concentrated flow erosion. The increased root biomass will furthermore increase the clod stability of soils, whereby the influence of plant roots will last longer.

In recent guidelines on erosion control in the Mediterranean region, new techniques have been suggested whereby it is suggested that vegetation is planted in "hotspots" or zones in the landscape where runoff occurs. Once remediation techniques have been carried out in these zones, degradation both at the site of the hotspot, and also off-site will be reduced (Recondes 2007). Examples of hotspots include river and gully banks and abandoned agricultural terraces. If planted with appropriate vegetation, the amount of water transmitted downslope will be decreased and sediment transport could be arrested. However, it is important to know which kind of vegetation is most useful for controlling erosion, depending on the type of hotspot and processes occurring there. Suggestions for vegetation strategies and typical plant species are given in Table 6-5.

Table 6-5. Potential vegetation strategies and plant species which could be applied to erosion hotspots in a Mediterranean region (after Recondes 2007).

Land type	Plants to be used for erosion control in specific zones
Reforested land	Vegetation should be planted where rills and gullies originate e.g. collapsing terraces; terraces not perpendicular to the slope direction and between rows if trees are planted in lines. On side banks, which are difficult and expensive to reforest, structures can be used to trap soil seeds and nutrients (see Chapter 7). Species to be used include grasses (*Stipa tenacissima* and *Brachypodium retusum, Helictotrichon filifolium*) and shrubs (side bank: *Salsola genistoides* and on other hotspots: *Rosmarinus officinalis*, *Anthyllis cytisoides, Rhamnus lycioides* and *Pistacia lentiscus*).
Croplands	It is more effective to cover the soil during the rainy season. Cover crops of weeds, legumes and grass species can be grown throughout the field, limited to strips perpendicular to the slope or in buffer strips along the field border. To conserve water resources in the summer, crops can be removed by tillage in the spring.
Abandoned lands	In fields, a quick establishment of vegetation cover (perennial species) with a fast growth rate, good vegetation cover and the ability to improve soil properties should be used. Where gully and rill erosion are problematic, vegetation can be planted on spots where concentrated flow can be expected. Grasses can be used (*Lygeum spartum, Brachypodium retusum* and *Stipa tenacissima*) in combination with deeper rooted shrubs (*Anthyllis cytisoides, Atriplex halimus* or *Salsola genistoides*) on terrace walls.
Hillslopes and gullies	Grass stems reduce runoff velocity and grass roots increase topsoil resistance to concentrated flow erosion and can prevent movement of soil blocks by increasing soil cohesion. Grass buffer strips or grassed waterways on the downslope border of a field could include the species *Stipa tenacissima, Lygeum spartum, Helictotrichon filifolium.* On steep slopes, shrubs e.g. *Salsola genistoides* would be useful. *Brachypodium retusum* and reed species e.g. *Juncus acutus* could be planted to vegetate drainage lines whereas for stabilizing gully floors a combination of grasses (*Lygeum spartum, Stipa tenacissima, Brachypodium retusum*), deep rooted shrubs (*Salsola genistoides, Anthyllis cytisoides, Atriplex halimus*) or trees (*Tamarix canariensis*) should be considered.

Channels	Different types of hotspots exist in channels, therefore vegetation strategies should be adapted depending on the erosion type. Grasses e.g. *Lygeum spartum* can be used on fans and *Stipa tenacissima, Lygeum spartum* on valley walls along with tree species (*Tamarix canariensis*). For larger tributaries/channels, consider either trees/shrubs (fine substrate – *Tamarix canariensis*, coarse substrate – *Nerium oleander*) and grasses (*Lygeum spartum*). Where water accumulates, plant *Juncus acutus* and *Phragmites australis*.

2. PERSPECTIVES FOR FUTURE STUDIES

Few studies exist concerning the exact mechanism by which trees and shrubs with different types of root systems fail during a landslide or avalanche. Field studies should be carried out in a forest after a landslide/avalanche event, and the type of failure quantified. Therefore, it will be possible to clarify the current indications about which species and type of root system resists best mass movement. Numerical modelling of tree failure during such events is an alternative to field studies, but should be validated by observations of real events.

The concept of using vegetation in hotspot zones of erosion should be examined further and the methodology applied to other types of soil mass movement and in different geological situations. In choosing the appropriate vegetation for stabilizing difficult zones, plant growth and root system morphology should be studied, as different species have a variety of strategies for growing in areas where soil erosion occurs. If these strategies were better understood, this knowledge could be expanded and applied to different situations.

In general, more information is needed on the ability of different plant species to fix soil in given situations. A database on root system architectural types would be of enormous help in identifying which species could be planted where, depending on the soil processes involved. Unfortunately, such a database is far from being developed, and for the moment, books by e.g. Köstler et al. (1968) and Kutschera and Lichtenegger (1997, 2002) must suffice.

3. REFERENCES

Abe K, Ziemer RR (1991) Effect of tree roots on shallow-seated landslides. USDA Forest Service, Gen. Tech. Rep. PSW-GTR-130

Alard D, Bance JF, Frileux PN (1994) Grassland vegetation as an indicator of the main agroecological factors in a rural landscape: consequences for biodiversity and wildlife conservation in central Normandy (France). J Env Manage 42:91-109

Barraclough PB, Wier AH, Kuhlmann H (1991) Factors affecting the growth and distribution of winter wheat roots under UK field conditions. In: McMichael BL, Persson H (eds) Plant Roots and their Environment, Proceedings of an ISRR Symposium, August 21–26 1988, Uppsala, Sweden. Elsevier, Amsterdam, pp 410-417

Bédécarrats A (1991) Dynamique des enherbements des pistes de ski en Savoie et leur gestion pastorale. In: IVth International Rangeland Congress, Association Française de Pastoralisme, Montpellier, France, pp 77-80

Bochet E, Poesen J, Rubio JL (2006) Runoff and soil loss under individual plants of a semi-arid Mediterranean shrubland: influence of plant morphology and rainfall intensity. Earth Surf Proc Land 31:536-549

Bouchon J (1987) Etat de la recherché relative aux dégâts forestiers dus aux tempêtes. Revue Forestière Française XXXIX 4:301-312

Brauner M, Weinmeister W, Agner P, Vospernik S, Hoesle B (2005) Forest management decision support for evaluating forest protection effects against rockfall. For Ecol Manage 207:75-85

Bűsgen M, Munch E, Thomson T (1929) The structure and life of forest trees. Chapman and Hall, London

Cammeraat LH, Van Beek R, Kooijman A (2005) Vegetation succession and its consequences for slope stability in SE Spain. Plant Soil 278:135-147

Coppin NJ, Richards IJ (1990) Use of Vegetation in Civil Engineering. CIRIA, Butterworths, London

Couturier DE, Ripley EA (1973) Rainfall interception in mixed grass prairie. Can J Plant Sci 53:659-663

Danjon F, Barker DH, Drexhage M, Stokes A (2007) Using 3D plant root architecture in models of shallow slope stability. Ann Bot-London, in press

De Baets S, Poesen J, Knapen A, Barberá GG, Navarro JA (2007) Root characteristics of representative Mediterranean plant species and their erosion-reducing potential during concentrated runoff. Plant Soil 294:169-183

Di Iorio A, Lasserre B, Scippa GS, Chiatante D (2005) Root system architecture of *Quercus pubescens* trees growing on different sloping conditions. Ann Bot-Lond 95:351-361

Dissmeyer GE, Foster GR (1985) Modifying the universal soil loss equation for forest land. In: El-Swaify SA, Moldenhauer WC, Lo A (eds) Soil Erosion and Conservation. Soil Conservation Society of America, Ankeny, pp 480-495

Dorren LKA, Berger F, le Hir C, Mermin E, Tardif P (2005) Mechanisms, effects and management implications of rockfall in forests. For Ecol Manage 215:183-195

Dorren LKA, Berger F (2006) Stem breakage of trees and energy dissipation during rockfall impacts. Tree Physiol 26:63-71

Eis S (1978) Natural root forms of western conifers. In: Symposium on Root Form of Planted Trees. Victoria, BC, Canada, pp 23-27

Forman R, Godron M (1986) Landscape Ecology. Wiley, Chichester, U.K.

Gray DH, Sotir RB (1996) Biotechnical and Soil Bioengineering Slope Stabilization: A Practical Guide for Erosion Control. Wiley & Sons, Inc., New York

Gyssels G, Poesen J, Nachtergaele J, Govers G (2002) The impact of sowing density of small grains on rill and gully erosion in concentrated flow zones. Soil Tillage Res 64:189-201

Gyssels G, Poesen J, Bochet E, Li Y (2005) Impact of plant roots on the resistance of soils to erosion by water: a review. Prog Phys Geog 29:189-217

Hartman R, De Boodt M (1974) The influence of the moisture content, texture and organic-matter on the aggregation of sandy and loamy soils. Geoderma 11:53-62

Henderson R, Ford ED, Deans JD, Renshaw E (1983) Morphology of the structural root system of Sitka Spruce. 1: Analysis and quantitative description. Forestry 56:121-135

Heumader J (2007) Revegetation on steep slopes and in subalpine areas using biennial cover plants: a review of Huter's technique. In: Stokes A, Spanos I, Norris JE, Cammeraat LH (eds) Eco- and Ground Bio-Engineering: The Use of Vegetation to Improve Slope Stability. Developments in Plant and Soil Sciences. Springer, Dordrecht, Netherlands, pp 427-438

Köstler JN, Brückner E, Bibelriether H (1968) Die Wurzeln der Waldbäume. Verlag Paul Parey, Hamburg & Berlin

Kupferschmid Albisetti AD, Brang P, Schonenberger W, Bugmann H (2003) Decay of *Picea abies* snag stands on steep mountain slopes. For Chron 79:247-252

Kutschera L, Lichtenegger E (1997) Wurzeln. Bewurzelung von Pflanzen in verschiedenen Lebensräumen. Stapfia 49, Land Oberösterreich, OÖ. Landesmuseum, Linz

Kutschera L, Lichtenegger E (2002) Wurzelatlas mitteleuropäischer Waldbäume und Sträucher. Leopold Stocker Verlag, Graz-Stuttgart

Lammeranner W, Rauch HP, Laaha G (2005) Implementation and monitoring of soil bioengineering measures at a landslide in the Middle Mountains of Nepal. Plant Soil 278:159-170

Mattia C, Bischetti GB, Gentile F (2005) Biotechnical characteristics of root systems of typical Mediterranean species. Plant Soil 278:23-32

Mickovski SB, Stokes A, van Beek LPH (2005) A decision support tool for windthrow hazard assessment and prevention. For Ecol Manage 216:64-76

Muller S, Dutoit T, Alard D, Grévilliot F (1998) Restoration and rehabilitation of species-rich grassland ecosystems in France: a review. Rest Ecol 6:94-101

Myers RD (1993) Slope Stabilization and Erosion Control Using Vegetation: A Manual of Practice for Coastal Bluff. Document number 30, Washington State – Department of Ecology, Ecology publications

Preston NJ, Crozier MJ (1999) Resistance to shallow landslide failure through root-derived cohesion in East Coast Hill Country soils, North Island, New Zealand. Earth Surf Proc Land 24:665-675

Prosser IP, Dietrich WE, Stevenson J (1995) Flow resistance and sediment transport by concentrated overland flow in a grassland valley. Geomorphology 13:71-86

Recondes (2007) Combating Land Degradation by Minimal Intervention: The Connectivity Reduction Approach, RECONDES team (eds), Published by University of Portsmouth, Portsmouth, UK, http://www.port.ac.uk/research/recondes/

Reid BJ, Goss MJ (1987) Effect of living roots of different plant species on the aggregate stability of two arable soils. J Soil Sci 32:521-541

Reubens B, Poesen J, Danjon F, Geudens G, Muys B (2007) The role of fine and coarse roots in shallow slope stability and soil erosion control with a focus on root system architecture: a review. Trees-Struct Func 4:385-402

Schiechtl HM (1980) Bioengineering for Land Reclamation and Conservation. University of Alberta Press, Edmonton, Alberta, Canada

Schmid T, Mueller U, Tognini F, Meyer J (2007) Comparison of revegetation techniques on alpine slopes prone to avalanches and erosion. In: Stokes A, Spanos I, Norris JE,

Cammeraat LH (eds) Eco- and Ground Bio-Engineering: The Use of Vegetation to Improve Slope Stability. Developments in Plant and Soil Sciences Volume 103, Springer Publishers, Dordrecht, pp 433-438

Schönenberger W, Noack A, Thee P (2005) Effect of timber removal from windthrow slopes on the risk of snow avalanches and rockfall. For Ecol Manage 213:197-208

Sidle RC, Pearce AJ, O'Loughlin CL (1985) Hillslope stability and land use. Am Geophys U. Water Res Monogr 11, pp 140

Stokes A (2002) The biomechanics of tree root anchorage. In: Waisel Y, Eshel A, Kafkaki U (eds) Plant Roots – The Hidden Half. Plenum Publishing, New York, pp 175-186

Stokes A (2006) Selecting tree species for rockfall protection forests. For Snow Land Res 80:77-86

Stokes A, Salin F, Kokutse AD, Berthier S, Jeannin H, Mochan S, Kokutse N, Dorren L, Abd.Ghani M, Fourcaud T (2005) Mechanical resistance of different tree species to rockfall in the French Alps. Plant Soil 278:107-117

Tasser E, Mader M, Tappeiner U (2003) Effects of land use in alpine grasslands on the probability of landslides. Basic Appl Ecol 4:271-280

Tengbeh GT (1993) The effect of grass roots on shear strength variations with moisture content. Soil Tech 6:387-295

Tsukamoto Y, Kusakabe O (1984) Vegetative influences on debris slide occurrences on steep slopes in Japan. In: Proc. Symposium Effects of Forest Land Use on Erosion and Slope Stability, Environment and policy Institute, Honolulu, Hawaii

Van Beek LPH, Wint J, Cammeraat LH, Edwards JP (2005) Observation and simulation of root reinforcement on abandoned Mediterranean slopes. Plant Soil 278:55-74

Van Noordwijk M, Brouwer G (1991) Review of quantitative root length data in agriculture. In: McMichael BL and Persson H (eds) Plant Roots and their Environment, Proceedings of an ISSR Symposium, August 21–26, 1988, Uppsala, Sweden, Elsevier Science Publishers, Amsterdam, pp 515-525

Weigel G, Shrestha RB, Meyer WP, Berg C (1987) Vegetative Soil Conservation Measures: A Field Manual. Vol. 1. Soil and Watershed Conservation Section Tansen, Palpa, Nepal

Wu TH (2007) Root reinforcement analyses and experiments. In: Stokes A, Spanos I, Norris JE, Cammeraat LH (eds) Eco- and Ground Bio-Engineering: The Use of Vegetation to Improve Slope Stability. Developments in Plant and Soil Sciences. Springer, Dordrecht, Netherlands, pp 21-30

Wu TH, Watson AJ, El-Khouly MA (2004) Soil-root interaction and slope stability. In: Barker DH, Watson AJ, Sombatpanit B, Northcut B, Magliano AR (eds) Ground and Water Bioengineering for Erosion Control and Slope Stabilization. Science Publishers Inc. USA, pp 183-192

Chapter 7

ECOTECHNOLOGICAL SOLUTIONS FOR UNSTABLE SLOPES: GROUND BIO- AND ECO-ENGINEERING TECHNIQUES AND STRATEGIES

Vicente Andreu[1], Hayfa Khuder[2], Slobodan B. Mickovski[3], Ioannis A. Spanos[4], Joanne E. Norris[5], Luuk K.A. Dorren[6], Bruce C. Nicoll[7], Alexis Achim[8], José Luís Rubio[1], Luc Jouneau[9], Frédéric Berger[6]

[1] Centro de Investigaciones sobre Desertificacion-CIDE, Cami de la Marjal, s/n, 46470 Albal (Valencia), Spain, [2] US2B, Université Bordeaux I, 33612 Cestas Cedex, France, [3] Jacobs UK Ltd., Glasgow G2 7HX, U.K., [4] National Agricultural Research Foundation, Forest Research Institute (FRI), 570 06 Vassilika, Thessaloniki, Greece, [5] Halcrow Group Ltd., Endeavour House, Forder Way, Cygnet Park, Hampton, Peterborough, U.K., [6] Cemagref Grenoble, 2, Rue de la Papeterie, Bp 76, St. Martin d'Hères Cedex, France, [7] Forest Research, Northern Research Station, Roslin, Midlothian, EH25 9SY, U.K., [8] Faculté de Foresterie et de Géomatique, Université Laval, Québec, G1K 7P4, Canada, [9] INRA, Domaine de Vilvert, 78352 Jouy-en-Josas cedex, France

Abstract: *For centuries vegetation has been used to prevent and control the effects of erosion and mass wasting processes. Techniques have developed continuously until now, when the increased environmental awareness of society has resulted in them being used as key tools in landscape conservation. The need for environmentally friendly techniques to mitigate the problems generated by soil instability (mass movements, rockfall, landslides, etc.) and the incidence of erosion have provoked the appearance in recent years of two different ecotechnological concepts: ground bio-engineering and eco-engineering. Both concepts are complementary, sometimes controversial, and have in common the use of biological materials (live and inert plant materials) as main and essential tools. In this Chapter, an updated and complete review of the different ground bio- and eco-engineering techniques in use is presented. The possible advantages and drawbacks of their application with regard to different degradation factors and processes are presented and future perspectives discussed. From the simplest methods such as seeding, mulching or planting, to the most complex ones that integrate different engineering techniques using very different materials (live cribwalls, vegetated gabions, etc.), we describe the uses of vegetation for increasing slope stability and restoring and preserving degraded land. The use of eco-engineering techniques against rockfall and windthrow, relevant problems in many European mountainous areas have also been considered. Finally, the possibilities of combining both eco- and bio-engineering techniques are described.*

J.E. Norris et al. (eds.), Slope stability and erosion control: Ecotechnological solutions, 211–275.

Key words: management strategies, slope stabilization, rockfall, windthrow, erosion control, restoration techniques, plant species, environmental conservation, forest fires

1. INTRODUCTION

Live vegetation and inert plant materials have been used for erosion control and also to stabilize and restore degraded slopes and river banks for several centuries, but generally using local knowledge and lacking specific scientific criteria. The first reference to this kind of engineering work was in 28BC in China (Redfield 2000). In ancient Greece, Sophocles warned against the intensive farming of olive trees. Although olive trees possess deep taproots, few surface roots exist to hold the topsoil in place (Stokes et al. 2004). The Roman writer Pliny stressed the importance of ditching and terracing slopes to control erosion, as early as the 1st century AD. In the 16th century, some cases of the use of willow plantings to control and stabilize slopes to prevent mass movements and erosion have also been reported (Lewis 2000).

In the last few decades, due to the increasing interest in environmental restoration and conservation, together with the implementation of ecotechnological solutions, the development of ground bio- and eco-engineering techniques has increased enormously. It must be remembered that vegetation slope interactions are very complex, difficult to quantify and to model, therefore any study must be tackled from an interdisciplinary approach, involving forest scientists, ecologists, geomorphologists, pedologists, geologists and engineers.

Both ground bio- and eco-engineering techniques have in common the use of biological materials, mainly plants and vegetation, as essential tools. Therefore, in many cases they can be used complementarily but this approach requires a careful appraisal, and the selection of species should be made carefully by considering the criteria given in Chapter 6.

This chapter highlights the advantages and disadvantages of the established ground bio-engineering techniques, introduces new strategies for protecting forests from substantial erosion damage, windthrow and rockfall, and finally reports on how both ground bio- and eco-engineering techniques can be used in combination to promote soil stability and land regeneration.

2. GROUND BIO-ENGINEERING TECHNIQUES

Slopes that are potentially suitable for ground bio-engineering require a careful choice of the particular ground bio-engineering technique. New slopes (e.g. embankments or cuttings) or slopes that have undergone land use change (e.g. terraces) may require planting, reforestation or seeding with appropriate species. The advantages and disadvantages and methods of application are described in Tables 7-1 and 7-2. Table 7-3 lists the most used ground bio-engineering techniques and their possible application in mitigation of some instability phenomena (see Coppin and Richards 1990; Gray and Sotir 1996; Schiechtl and Stern 1996 for further information). Existing slopes that are either unstable from soil erosion or from shallow slope failure may be suitable for the ground bio-engineering techniques that are described in Tables 7-4 to 7-35.

Some important considerations when establishing vegetation on slopes are:

- Loss of vegetation leaves the slope vulnerable to runoff, erosion and sedimentation. Furthermore it enhances weed growth, degrades habitats and decreases forest regeneration. In order to combat the consequences of loss of vegetation on slopes, revegetation strategies can be adopted, in which seeding and planting will be major treatments.
- The choice of the best applicable treatment depends on the nature of vegetation loss (forest fire and its intensity, sylvicultural operations e.g. clearcuts, etc), slope type and inclination, proximity to drainage, possibility for weed spread and the management objectives.
- In semi-arid conditions, like those characteristics of Mediterranean environments, the plantation technique to use, the place and the hole design (for runoff collecting) should be selected very carefully. In the same way, the season for planting must be chosen, being preferably in autumn, but not in the period of hydrological deficit (spring or summer).
- The vegetation along the edge of the top of the slope serves as a protective buffer for the slope face. If possible, a greenbelt which would provide a buffer between the slope face and residential constructions should be maintained or re-established.
- Vegetation should be established on patchy and barren slope faces or terraces to reduce erosion (see Chapter 6). Various species and mixtures of them can be planted on slope faces and expected to succeed in this rather severe environment. These include seed mixtures of grasses and legumes and a range of shrubs and minor trees.
- Large trees should be used on the face of slopes sparingly and with caution. These trees could collapse because of undermining of the root system by erosion or by windthrow, large volumes of earth can be disturbed by the tree roots when they are pulled away from the slope. The resulting large, bare areas are opened to further erosion, which may endanger adjacent land and vegetation. If the trees become unstable, they should be cut or coppiced before they fall. Root systems should be left intact to bind the soil for a short period of time while new live, well-rooted vegetation is established. Planting new vegetation prior to felling a tree would be advantageous to the slope protection program.
- In those situations where the bottom of a slope is susceptible to frequent or periodic water erosion, e.g., at the coast, vegetation alone will not be

adequate as an erosion control tool. In such cases a form of structural toe protection may also be required. If the toe is not subject to coastal marine erosive forces, trees and woody shrubs can be useful in resisting upland landsliding and tolerating the dynamic changes in the coastal shore system. Vegetation at the slope toe can sometimes help reduce marine erosion to manageable levels.

Table 7-1. Planting and reforestation techniques

Application	Advantages	Disadvantages	Effectiveness
On slopes with maximum inclination of 1:1.5 (V:H) On low banks and marine estuaries	Fast action program for specific slope areas Higher plant survival Minimum slope disturbance when using planting in holes	Does not solve some erosion problems (gully erosion) Container grown plants might be expensive Hard to install in some mulching systems Has to be performed in dormant season (late autumn or early spring) and requires watering	Plant root systems penetrate into the lower soil horizons and stabilize the soil Plant roots can subsequently drain the slope by using underground water for survival
Material	**Diagram**		
Plants installed in groups or at specific distances and then pruned Plant selection is dependent on site conditions and erosion problems Structural diversity in plant selection (trees/shrubs with ground cover) is effective Planting should be done during dormancy and when water is available	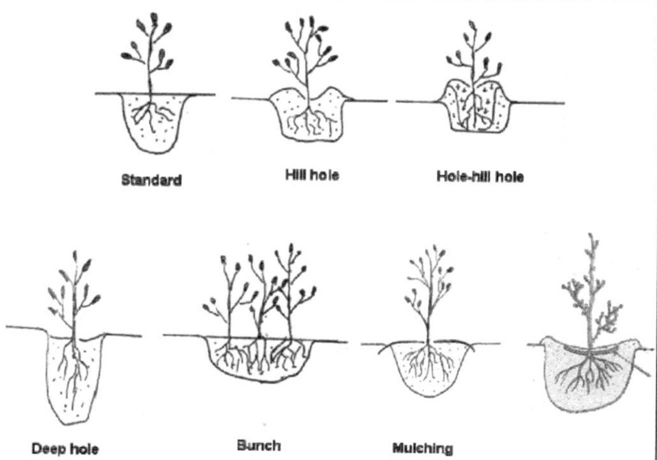		

Table 7-2. Seeding techniques

Application	Advantages	Disadvantages	Effectiveness
On mild slopes, in small-scale areas affected by erosion processes Usually applied in combination with other planting techniques	Quick application Low cost of materials Compatible with many slope situations	Does not readily self-repair eroded slope areas, and should not be applied alone in highly eroded areas or for shallow seated landslide stabilization Seed needs to be mulched immediately to avoid it washing/blowing away, or the action of any fauna, mainly rodents Soil needs to be kept moist	Creates a shallow fibrous rooting zone in the uppermost 0.30 m of the soil which binds the surface soil particles and protects soil surface from runoff, wind and freeze-thaw erosive processes

Material	Observations
Grass, forb and woody plant seed mixes are sown directly or hydro-seeded Perennial grasses and forbs (for long term cover but slower to establish) for severely and moderately disturbed sites which are less than 15 m to a drainage channel Annual ryegrass and small grains should be seeded on moderately disturbed slopes of 15% and more inclination Seeding should be done in late autumn or early spring, or in the case of wildfires, immediately after the fire when the soil surface has lost to some degree its vegetation cover	Loss of vegetation leaves the slopes vulnerable to increased runoff, erosion, and sedimentation. Furthermore, it enhances weed growth, degrades habitats and decreases forest regeneration. In order to combat the consequences of vegetation on slopes, a revegetation strategy can be adopted in which seeding and planting will be major treatments. The choice of best applicable treatment depends on the nature of vegetation loss (forest fire and its intensity, silvicultural operations like clearcuts, etc), slope type and inclination, proximity to drainages, possibility for weed spread, climate conditions and the management objectives. In Mediterranean conditions, the use of this technique is closely dependent on the soil water regime. Slopes that suffered severe or moderate vegetation loss e.g. after a fire, in some cases, should be reseeded to minimise the likelihood of erosion and sediment movement downslope. For slopes suffering from light vegetation loss, reseeding is not necessary since they can recover quickly. Native species should be used where the re-establishment of the native plant community is the primary objective. Introduced species should be used when stabilization and resource protection are main objectives. A mixture of native and introduced species is not recommended since the introduced species might hinder the establishment of the natural species.

Table 7-3. Ground bio-engineering techniques and their possible application in mitigation of some instability phenomena on slopes

Bio-engineering technique	Shallow slides	Moderate mass movement	Mud slides	Debris creep/flow	Soil creep	Seepage erosion	Overland flow erosion	Splash erosion	Rill erosion	Gully erosion	Pipe erosion	Streamflow erosion	Flood inundation	Avalanche
Branch layering in gullies			Y	Y						Y				
Branchpacking	Y					Y	Y	Y	Y	Y				
Brush mattress construction						Y	Y	Y				Y		
Brush wattles	Y*					Y	Y	Y	Y			Y		
Brushlayer construction	Y	Y	Y	P	Y	Y	Y					Y		
Contour log terraces	Y		Y	Y		Y			Y	Y	Y	Y		
Contouring, sloping, regrading			Y					Y						
Cordon construction	Y		Y	Y	Y									
Crib-wall construction with branchlayering	Y		P	P	P									
Earth berm water bars						Y	Y		Y	Y		Y		
Furrowing, contour scarification					Y	Y		Y				Y		
Grassed waterways							Y	Y					Y	
Gravel drains							Y		Y	Y		Y	Y	
Groove construction						Y	Y	Y	Y				Y	
Hedge brushlayer construction	Y		Y		Y								Y	
Hedge layer construction	Y		Y		Y								Y	
Live crib walls (concrete and prefabricated elements)	Y		Y	Y	Y									
Live fascine drains							Y		Y	Y			Y	
Live pole drains							Y		Y	Y			Y	
Live shoring of open water canals							Y		Y	Y			Y	
Live slope gratings						Y	Y	Y	Y	Y				
Live staking/live fascine	Y		Y		Y		Y	Y				Y		
Matchsticks							Y	Y	Y	Y				
Mulching	P						Y	Y					Y	
Placing of cuttings and wall-joint planting	Y			P			Y							Y
Silt fences	Y		Y	Y										
Slope drainage using phreatophytes	Y													
Sodding or turfing	Y*					Y		Y	Y	Y				
Straw bale check dams	Y		Y	Y										
Vegetated gabions	Y			P					Y	Y		Y		
Vegetated geogrids		Y	Y			Y						Y		
Vegetated palisade and pole construction	Y				P					Y		Y		
Vegetated stone walls and rock piles	Y				P									
Wattle fences	Y		Y				Y	Y	Y		Y			

* subject to successful rooting; Y successful bio-engineering technique; P not a proven successful technique.

Table 7-4. Branch layering in gullies

Application	**Advantages**	**Disadvantages**	**Effectiveness**
For repairing of shallow gullies (no deeper than 3 m and no wider than 8 m)	Provides continued effectiveness through the use of live plant material	Slightly more expensive than dead branch layering of gullies Cannot cope with continuous flow Cannot be applied if severe bed load and shoulder movement with significant deposition is expected	Live branches root and secure the gully bed. Well rooted branches can withstand temporary flooding Silt should not cover more than a third of the annual growth of the branch
Material	**Diagram**		
Long and strong live branches of rooting plants (for gullies deeper than 1.5 m, very bushy branches can be used) Cross beams placed at a distance of 2 m, with length and thickness depending on the gully	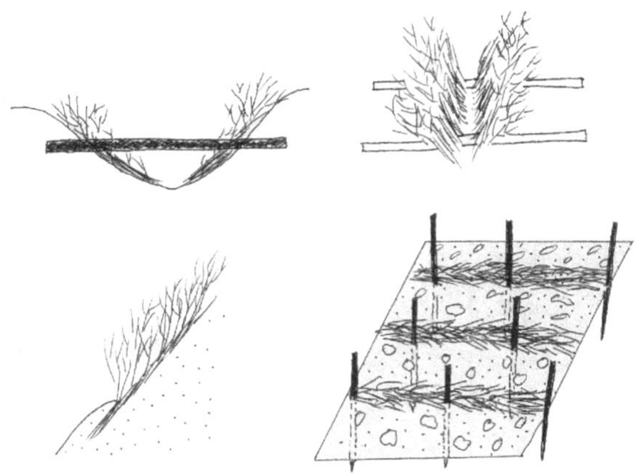		

Table 7-5. Branchpacking

Application	**Advantages**	**Disadvantages**	**Effectiveness**
For repairing of small localized slumps and holes (0.005 to 0.01 m in width and depth) in stream banks	Effective Inexpensive Provides immediate soil reinforcement Rapidly establishes a vegetated stream bank	Not effective for slumps and holes wider and deeper than 1.0 m	Produces a filter barrier that prevents erosion and scouring from stream bank or over bank flow Live branches serve as tensile inclusions for reinforcement once installed As plants begin to grow, the system becomes more effective in retarding runoff and reducing surface erosion Trapped sediment refills the localized slumps or hole, while roots spread throughout the backfill and surrounding earth to form a unified mass
Material	**Diagram**		
Wooden stakes 1.5 to 2.0 m long, 0.05 x 0.10 m in cross section driven to 1.0 to 1.2 m into the undisturbed soil Live branches 1.5 to 5 cm in diameter inserted between compacted backfill Toe bank protection of large stones and geotextiles may be required at the toe of the slope in stream banks	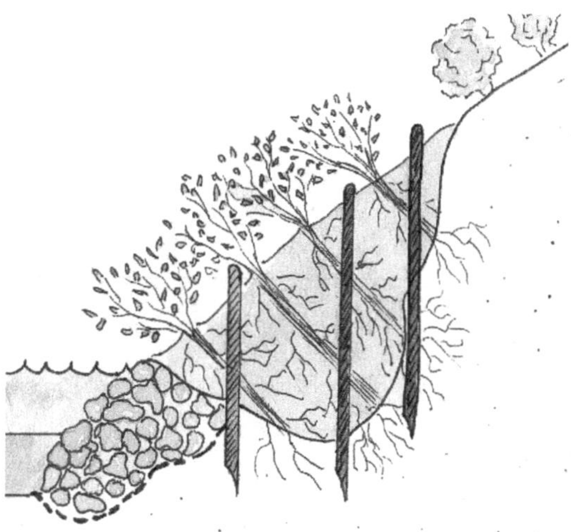		

Table 7-6. Brush mattress construction

Application	Advantages	Disadvantages	Effectiveness
Surface protection Water/wind/wave erosion protection Protection of water channel banks against flowing water Repairing damaged areas	Immediate effectiveness even before the plants root Dense root and thicket development	Much material and labor is needed The effect of soil stabilization is lower than the one of brush layers Thinning may be required	Immediate cover and protection Roots can penetrate deeply if the soil is dry and permeable Permanent effect with live materials Possibility for the climax vegetation to establish itself quickly
Material	**Diagram**		
Long (>1.5 m), straight branches which root easily Smooth branches (5 kg/m^2) Bushy branches (5 to 10 kg/m^2) Live and dead material can be mixed 20-50 branches per meter length of the construction	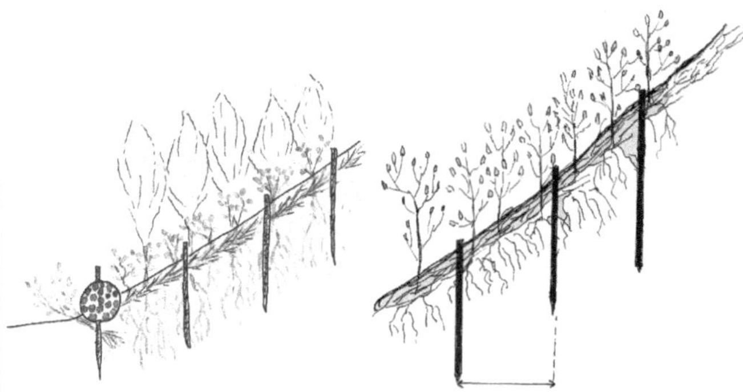		

Table 7-7. Brush wattles (slope fascines)

Application	Advantages	Disadvantages	Effectiveness
For cut slopes in deep and soft sand In low altitudes with good growth conditions Areas where live branches are available and where fast growth can be expected	Very fast construction Simple Little soil disturbance	Lateral spreading branches cannot be used The system is susceptible to rockfall	Slope stabilization is provided by shading the soil and penetration of the roots
Material	**Diagram**		
Long and straight branches of live woody plants Each fascine contains 5 branches with diameter of around 0.01 m, and pegs (>0.60 m/m) and are held in place by either wooden stakes, live fascines, gabion nets or large stone blocks (as illustrated from left to right)	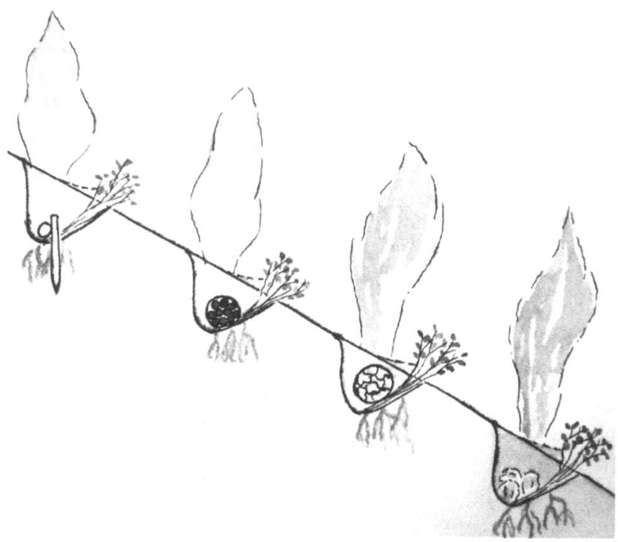		

Table 7-8. Contour log terraces/barriers

Application	**Advantages**	**Disadvantages**	**Effectiveness**
Post-fire treatment providing obstacle to runoff from heavy rainstorms On slopes with an angle that varies from 31-50° On burned slopes where there are a number of dead trees that have little or no economic value	Local materials used Inexpensive Development of soil barriers with time Allows the establishment of vegetation	Cannot be used on steep slopes and heavy machinery must be avoided Enough trees must be felled to create a barrier that interrupts the movement of water and sediment downslope Little or no effect achieved if the logs are not in contact with the soil	Logs are placed in an alternating scheme so the runoff no longer has a straight down slope path to follow, reducing its kinetic energy. The water is forced to meander back and forth between logs, reducing the velocity and energy of the runoff, and giving water time to infiltrate into the soil.

Material	**Diagram**
Dead trees are felled, limbed, and placed on the contour perpendicular to the direction of the slope. The logs should be bedded into the soil for its entire length and backfilled with soil so water cannot run underneath; backfill should be trampled down. Logs should be secured from rolling by driving stakes on the downhill side.	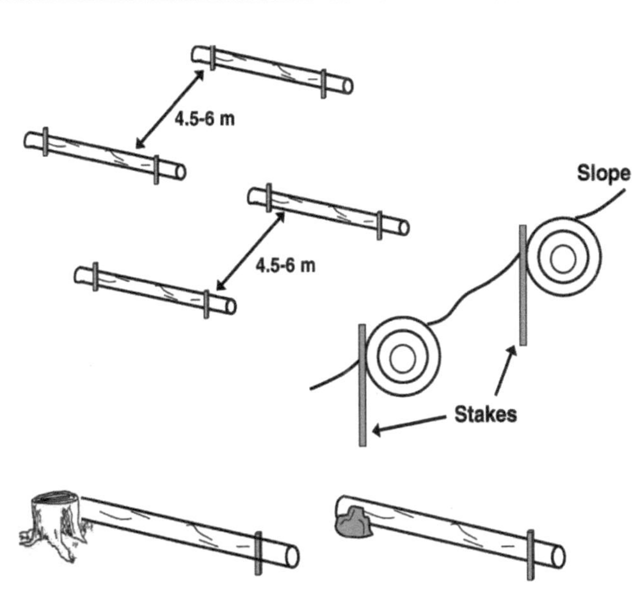

Table 7-9. Contouring, sloping, regrading

Application	Advantages	Disadvantages	Effectiveness
Low slopes with enough space at the top to allow access	Slopes can be left steeper than their natural angle of inclination	Neither economically feasible nor technically desirable for an individual property owner	Produces an ideal form of a slope without sharp edges, especially at the top and the toe

Materials	Description
Most commonly used method is regrading with effective earth moving machines, but only on sites where there is no problem with deposition of the excess material Water pressure (underwashing or inducing artificial slides) applied from the toe to the crest is a more viable option if local conditions allow it	Proper rounding off will cover every difference between the cut and the natural landscape Grading the slopes to an inclination of 1:3 (V:H) or flatter is ideal because these slopes can be prepared and planted with wheeled vehicles Blasting, drilling and jackhammering usually are expensive and they do not produce desired results

Table 7-10. Cordon construction

Application	Advantages	Disadvantages	Effectiveness
Moist slopes with clayey soils, heavy clay soils, limestone soils, mica slate soils, soil containing schistose material Dry slopes Couturier method is particularly effective for reafforestation of dry slopes.	**Couturier method** Excellent for water retention in dry climatic zones **Praxl method** Stabilizes suitable slopes Offers high resistance to slides and slippages Improves the aeration of the plant roots	**Couturier method** Should not be used on slopes prone to slipping Offers high risk of water impoundment **Praxl method** Has high labor and material costs Might cause damage in the surrounding shrub or forest areas More economical and effective methods exist (hedge brush layer, brush layer)	**Couturier method** Improves slope stability by retaining water and levelling out the planting beds **Praxl method** Strong branch overlay provides very good stabilisation of suitable slope sections Provides good root penetration
Materials	**Diagram**		
Couturier method the (right hand drawing) three rooted seedlings of trees or shrubs for every running meter, 2 to 5 cuttings at 0.10 m from the sloping ground surface **Praxl method** (left hand drawing) two posts 0.06 to 0.12 m in diameter 10 to 25 cuttings with a minimum length of 0.50 m between the two posts			

Table 7-11. Earth-berm water bars

Application	**Advantages**	**Disadvantages**	**Effectiveness**
On slopes after a high/very high intensity fire The local soils and the road/trial grade will dictate the spacing between the berms	Properly built earth-berm water bars are very effective in diverting water off roads, trails, and landings. They also limit undesirable traffic following closure.	Hard to drive over and may be difficult to maintain They do not work well for active traffic surfaces during most operations. Frozen soils and rock may limit their use. They require caution when blading to maintain the road	Channel water off roads and trails to avoid the creation of gullies Water bars are angled down slope to the outlet side and can divert water to a vegetated slope below or redirect it into a channel that will take it to a culvert

Material	**Diagram**
Berms of soil or embedded logs	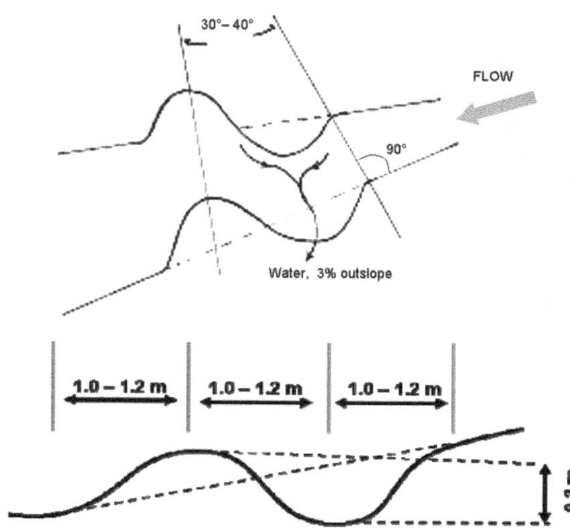

Table 7-12. Furrowing, contour scarification

Application	Advantages	Disadvantages	Effectiveness
In moderately to severely disturbed (burned) areas Burned upland areas with hydrophobic soil properties On slopes 0-30° to facilitate safe operation by machinery	Effective as a preparatory measure before vegetation seeding Multiple gains for reducing soil loss	Not to be used in swales, drainage ways, gullies or other areas of concentrated flow Requires usage of machinery	To break up the hydrophobic soil layer To aid in the establishment of vegetative cover from seed To reduce runoff velocity To increase infiltration To reduce erosion
Material	**Diagram**		
Small tractors, bull dozers or all-terrain vehicles equipped with a tool bar with tines, rippers or other scarification devices capable of loosening and mixing the soil to a depth of 0.05-0.10 m Can be done in strips 2-3 m wide spaced uniformly over the slope. The spacing between strips can be between 10 m for slopes with 20-30° inclination up to 50 m for slope inclinations less than 5°			

Table 7-13. Grassed waterways

Application	Advantages	Disadvantages	Effectiveness
For slope drainage For surface water drainage around the toe of a slope Road construction Regulation of the water drainage on ski runs Artificial fill slopes or earthworks	Effective immediately if sods are used Easy to check its functioning because it can be viewed from above Blends well into the landscape	Very difficult to establish on rocky slopes Cannot be used for gullies with a steady water flow	Effective for the channeling of surface water Sods act as a water pump in draining the slope, especially in waterlogged soils

Material	Description
Sods, reed sods, seed mats, hydro seeding material, pegs, hay, straw, wire or plastic netting, bitumen	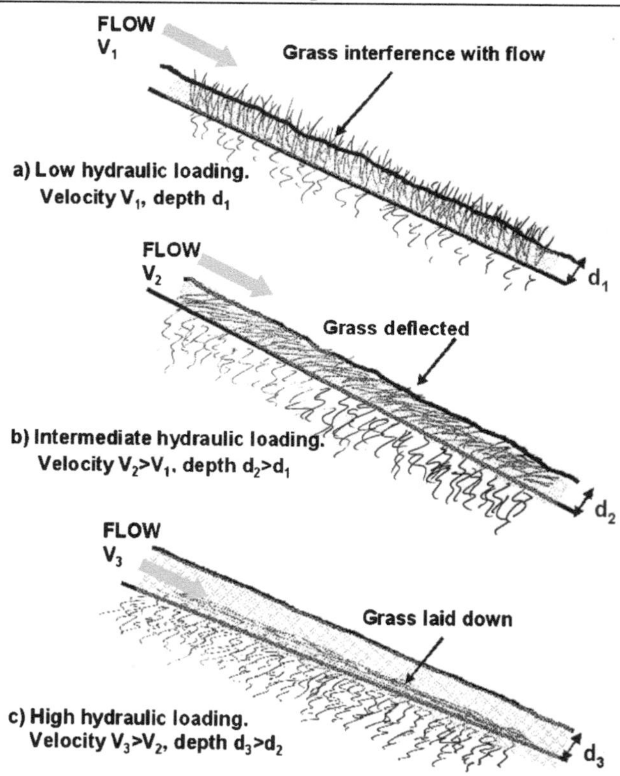 after Hewlett et al. 1987.

Table 7-14. Gravel drains

Application	**Advantages**	**Disadvantages**	**Effectiveness**
Instantaneous repair of slides Catching water layers at the toe of the slope Protection from frost damage	Simple system with permanent effectiveness More attractive than conventional engineering construction No maintenance required if far enough from roots	Only possible where machines are available and rocks or gravel are on hand Height is limited by vehicular access	Acts immediately as a support and a drain
Material	**Diagram**		
Rocks or gravel Branches of live woody plants (several meters long)	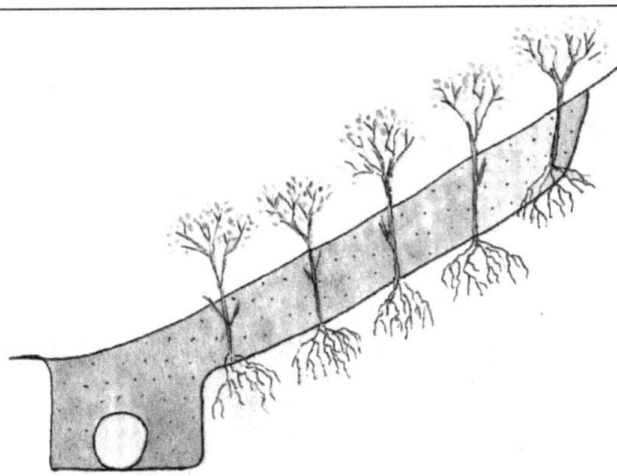		

Table 7-15. Hedge brush layer construction

Application	**Advantages**	**Disadvantages**	**Effectiveness**
Fill slopes (where danger of erosion, slides, rock fall exists) Dry slopes Riverbanks Water channel protection Steep slopes protection Slopes of rocky and loose material Cut slopes	Simple Heavily branched twigs can be used Less expensive Little loss of plants Low material demand In one operation two stages of vegetation community plant succession are established	Unsuitable for retaining topsoil	Best penetration effect of all stabilizing constructions. It starts immediately and increases with rooting. The microclimate improvement on the slope surface is effective. Gully erosion can be stopped if the brush layers are constructed on longitudinal strips of dead material. The inclusion of nitrogen-fixing plants will reduce soil nitrogen deficiency and will improve soil condition rapidly

Material	**Diagram**
Fill slope 1 or 2 rooted healthy plants (fast growing pioneer plants, several years old); 1.5-5.0 m spacing, inserted 2.0-5.0 m into the slope at a minimum gradient of 10% (left diagram) **Cut slope** 10 branches of live woody plants with all their side branches; 1.5-5.0 m spacing, inserted 0.5-2.0 m into the slope at a minimum gradient of 10% (right diagram)	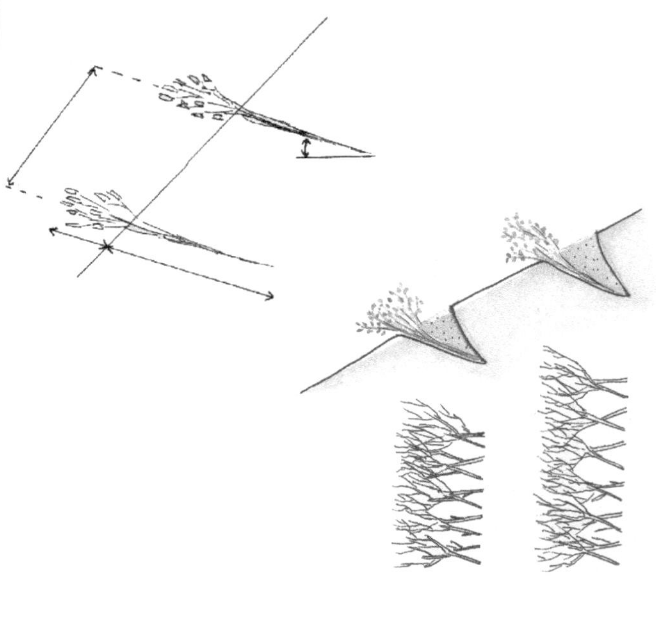

Table 7-16. Hedge layer construction

Application	Advantages	Disadvantages	Effectiveness
Good soils Fertile loess and gravel soils Sandy and clayey soils Areas where there is no material available	Enables creation of a forest plant community with closed canopy without planting pioneer species	Large quantity of plants required Very high cost	Soil stabilization begins immediately after construction, but hedge layers are most effective in long term Soil penetration is good among with soil improvement, soil activation, and shading Woody plants that will create the climax community should be used

Material	Diagram
Rooted woody plants (2 to 4 year-old) that are fast growing and very resistant 5 to 20 plants per running meter upslope spacing between 1.0 m and 3.0 m apart, inserted at a minimum gradient of 10%	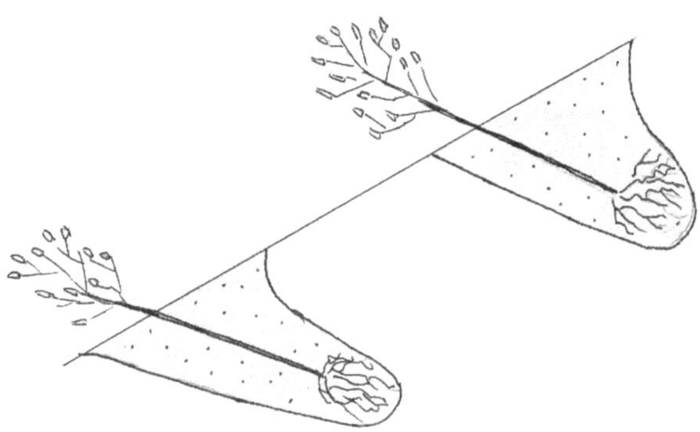

Table 7-17. Live cribwalls (concrete and prefabricated elements)

Application	**Advantages**	**Disadvantages**	**Effectiveness**
Urgent repair of disaster stricken areas Repair of slides Shore or steam channel bank protection (instead of solid concrete walls) Stabilization of slopes, slope sections, toes of slopes, water channel beds	Provide excellent stability Fast and simple construction Suitable for urgent repair work after a disaster	Not very good for the landscape Construction has relatively high costs Very heavy materials are used	The rotting timber is replaced by the growing plants The established plants drain the slope effectively through transpiration

Material	**Diagram**
Branches of plants that will root from cuttings (10 branches per running meter of construction) A single live cribwall is illustrated	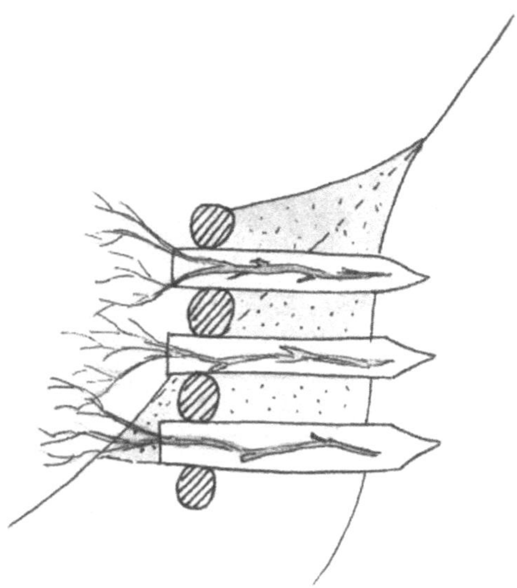

Table 7-18. Live cribwalls

Application	**Advantages**	**Disadvantages**	**Effectiveness**
Areas where a catastrophe (soil instability) has already occurred For stabilization of parts of slopes, water channels, and toes of slopes Reinforcement constructions for linear and/or spatial slope stabilization	Fast stabilization Short building period Can be constructed in a horizontal line Provide active drainage and the increase of the root systems' armouring effects	The lumber can lack durability	Plants drain the slopes very effectively through transpiration Single or double crib walls consisting of timber, concrete, metal or synthetic materials represent technical stabilization elements, whilst the simultaneous use of live plant material and branch inlays initialize the establishment of the vegetation
Material		**Diagram**	
Round or square timber (0.10-0.25 m in diameter, at 1.0-1.5 m spacing). Strong 1 m long branches from species that root easily (10 branches per running meter of construction) such as larch, silver fir, pine, oak, European chestnut or black locust A double live cribwall is illustrated	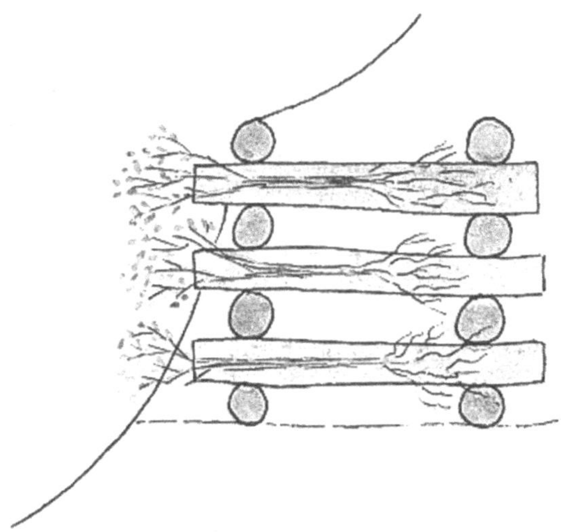		

Table 7-19. Live fascine drains

Application	**Advantages**	**Disadvantages**	**Effectiveness**
Slope drainage where the water is not too deep-seated Suitable for extensive surface area drainage	Simple Fast Less expensive and more attractive than conventional engineering construction	Construction only possible during the dormant season	The channeling effect of the longitudinal branches enables effective fascine drainage immediately after the placement Desiccates the area further by transpiration after the development of the roots
Material	**Diagram**		
Very long live branches tied together in bundles and staked in to the ground	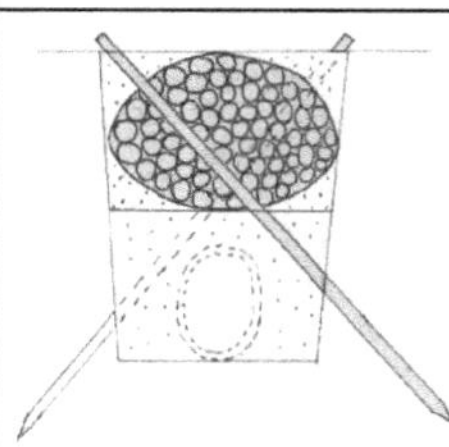 fascine bundle overlies gravel fill, staked with a live plug	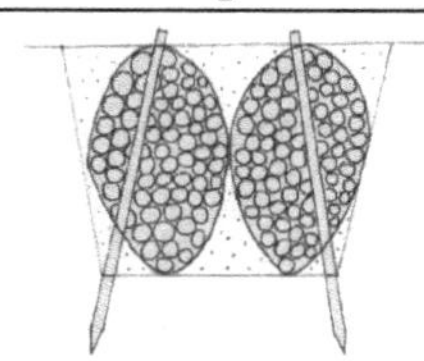 fascine bundles placed next to each other in a hole and staked with a live plug	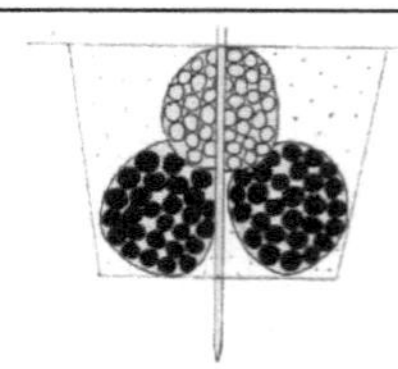 a live fascine bundle sits on top of two dead bundles and staked

Table 7-20. Live pole drains

Application	**Advantages**	**Disadvantages**	**Effectiveness**
Slope drainage where the water is not too deep-seated Suitable for extensive surface area drainage	Usually better growth is obtained than with fascine drains Cheaper than hard engineering construction	Higher cost (higher consumption of material that is difficult to obtain)	Only difference from fascine drains is the use of sturdy live branches (instead of slender ones) either loosely arranged in the ditch and secured with crossbeams or tied with pegs and secured with timber and covered with gravel
Material	**Diagram**		
Live poles (heavy and rigid branches or small trees) of 3-14 cm in diameter Dead material for the bottom of the ditch Live pegs or timber 0.8 m long which form the sides of the drain	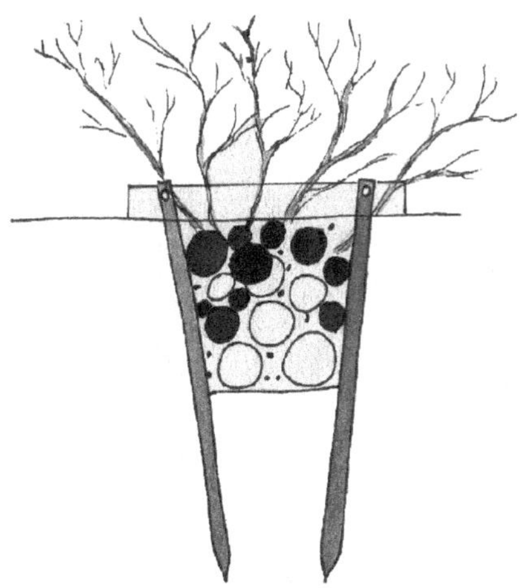		

Table 7-21. Live shoring of open water canals

Application	**Advantages**	**Disadvantages**	**Effectiveness**
Slope drainage Surface water drainage around the toe of a slope Road construction Water drainage regulation on ski runs Useful for temporary or continuous low water flow Where open drainage is required	Cheap to construct	Costs can be much higher if boards or plants are used for securing the walls or the bottom	The channeling effect of the longitudinal branches enables effective fascine drainage immediately after the placement Desiccates the area further by transpiration after root system development

Material	**Diagram**
Live branches or poles Live pegs 1 m long (left diagram) Boards can be used to secure bottom of potentially steady stoke flow channel (right diagram)	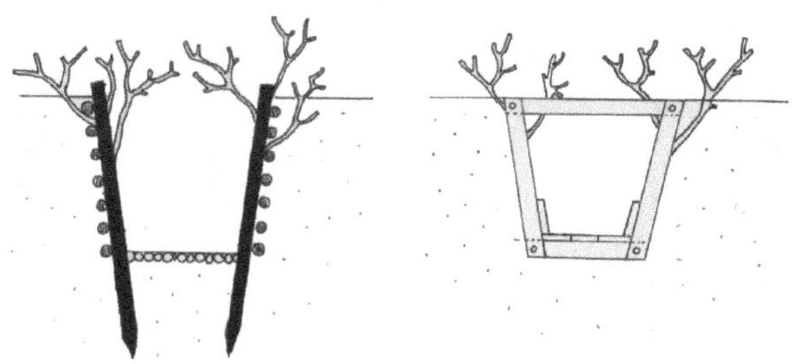

Table 7-22. Live slope gratings

Application	Advantages	Disadvantages	Effectiveness
Very steep slopes where angle cannot be reduced, with height of the gratings between 10 and 15 m Infrequently used method (sloping is preferred)	Immediate effectiveness Combinations and variations are possible	High labour costs	The live building material for the grating denotes that the entire protection system is alive and rooted in the slope at the same time, thus stabilizing and draining it
Material	**Diagram**		
Round or square timber, corresponding to the dimensions and the type of construction either nailed together or tied with wire, and clamped at the base	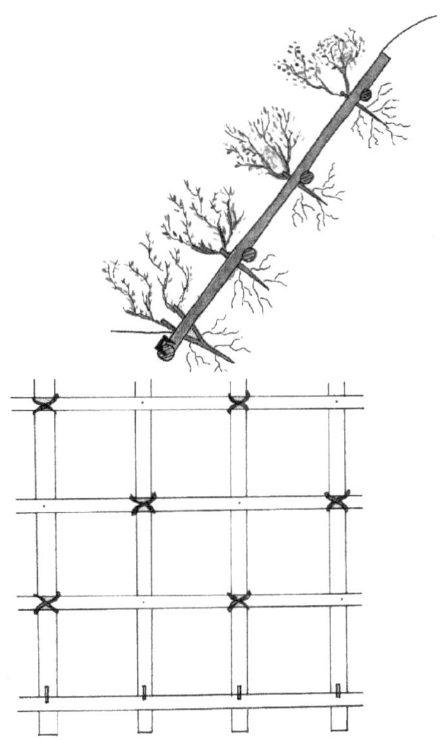		

Table 7-23. Live staking

Application	**Advantages**	**Disadvantages**	**Effectiveness**
Where single stem plantings will provide adequate plant cover, slope stability and fish habitat Can be applied on stable, irregular slope surfaces	Plentiful and inexpensive material Can be applied with minimum slope disturbance Helps in reducing slope soil moisture It may be combined with other revegetation techniques to anchor bundles, brush mats and erosion control fabric	Not a short term solution to slope instability problems Does not solve existing erosion problems Live stakes require moist soils, but watering is not required (although it can increase survival and promote plant growth)	Simple technique that installs a dormant cutting directly into the ground Occasional deep watering is more effective and encourages deeper rooting than frequent light watering

Material	**Diagram**
Several live stakes (0.25 to 0.65 m long, 0.005 to 0.015 m in diameter) from a dormant cutting should be buried upwards on a distance of 0.30 m to 1 m with only one or two buds left exposed out of the soil Water during the first 6 weeks after planting if the soil is dry	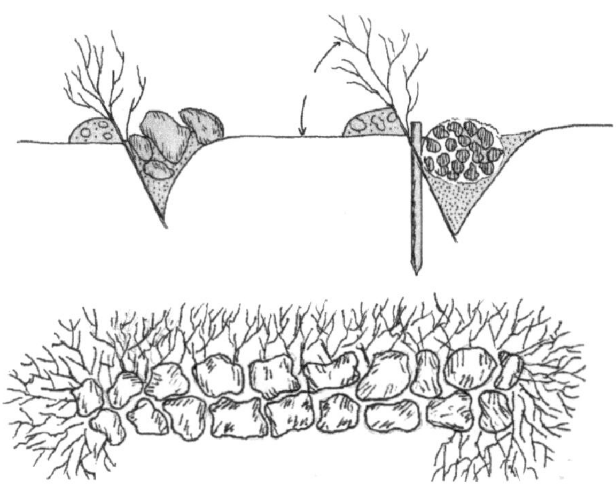

Table 7-24. Matchsticks, vertical mulching

Application	Advantages	Disadvantages	Effectiveness
On slopes with an angle between 0-30° after a medium or high intensity fire Large sandy areas	Perform very well in dry climates Cheap Does not leave permanent patterns on the landscape after removal Increases soil moisture storage >20%	Not effective on steep slopes Not applicable on slopes with rock face	Slowing water movement Provides open channels for water penetration into the deep soil Collecting the sediment, sand and stones moving downwards from the slopes Stopping soil erosion during heavy rainfall Provides both wind breaks to trap seeds and dust and shade and cover for seedlings

Material	Diagram
Branches, branchlets, thin boles, and the remnants of clear felling, stacked on the ground in horizontal lines, (on the uphill side of the stumps) should be arranged in rows The distance between these rows has been calculated to be 10-15 m depending on relief	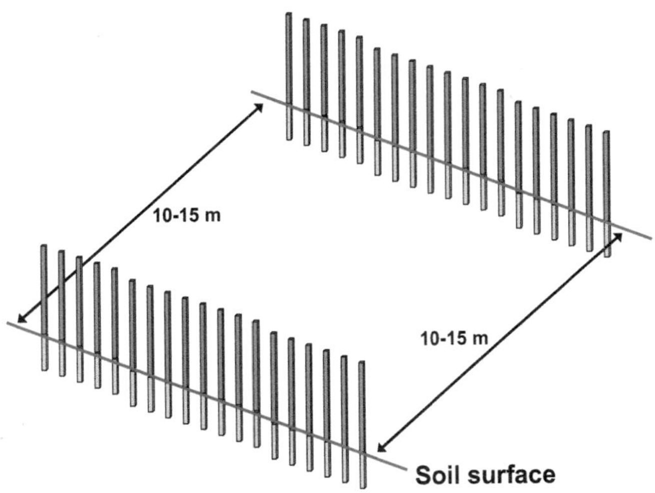

Material	Diagram
Their dimensions are, height 0.50-0.75 m, width 1.0-1.5 m and length from 1 to a thousand and more meters Several materials can be used for vertical mulch, including: broom corn, straw, brush and reeds. The best choice for a given site will depend on availability and cost of materials, project demand for aesthetics, integration of seeding and container planting, and severity of erosion and land stability problems.	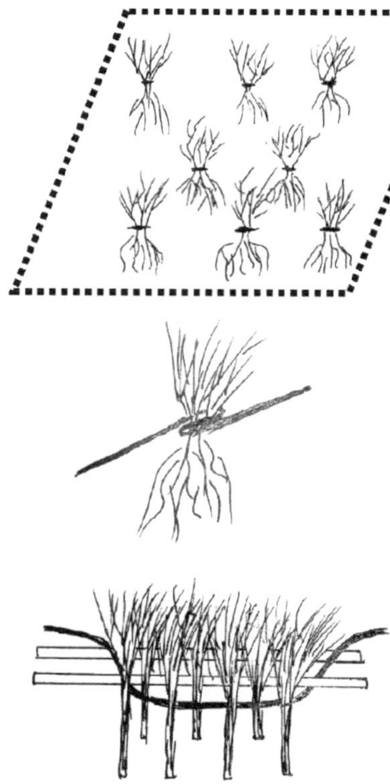

Table 7-25. Mulching

Application	Advantages	Disadvantages	Effectiveness
For protection of slope plantings On slopes with high erosion potential On slopes affected by forest fires On coastal slopes a mulch cover is necessary if vegetation is to be established from seed	Can be done quickly and at low cost even using mechanization It can be applied even on long and flat slopes Maintain soil moisture	Restricted to sites where there is an access for mechanization Limited to slopes with inclination less than 1:1 (V:H) (45°)	Protects against rain and wind while seeds are germinating Reduces loss of soil moisture during extended dry periods Reduces heaving (plant roots forced upward out of soil) of small plants as a result of alternate freezing and thawing

Material	Diagram
Hay or straw (250-500 g/m^2), bamboo, reed, jute netting, plastic netting (not recommended), manure or compost (not recommended), wood fiber or fiber matting Anchoring of the mulch can be provided with hand, roller or crimper punching, or alternatively with erosion control netting Must be punched into the soil or covered with erosion control netting	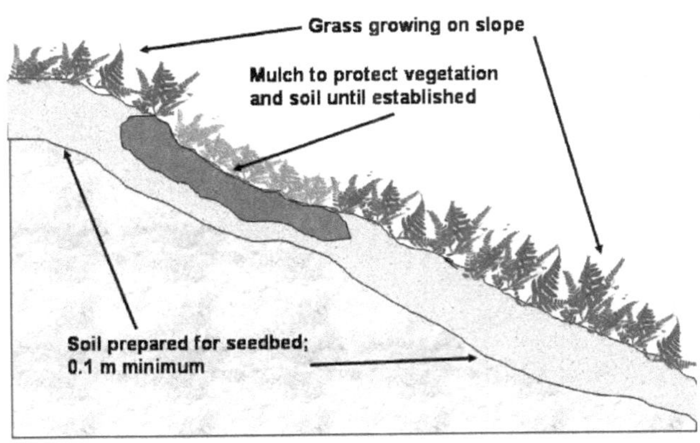

Table 7-26. Placing of cuttings and wall joint planting

Application	Advantages	Disadvantages	Effectiveness
Planting on moist slopes for controlling wind, water, and avalanche erosion Reinforces rock paving in earthworks and in avalanche protection constructions Available for vegetation on stone piles	Inexpensive Quick building Excellent effect along an entire area Rock paving enables the use of smaller otherwise unsuitable rocks	Stabilization does not start before the plants are rooted The operation is only possible during the period of dormancy	Soil stabilization and drainage strengthening achieved with plant roots Strengthens avalanche brake constructions, avalanche diversions, channel protection walls, or channeling walls Improves the microclimate The falling leaves protect the rock wall effectively
Material	**Diagram**		
1 to 2 year-old cuttings without branches (diameter 0.02-0.04 m, length 0.20-0.40 m) If the water supply or retention is poor the cutting should be 0.40-0.60 m long	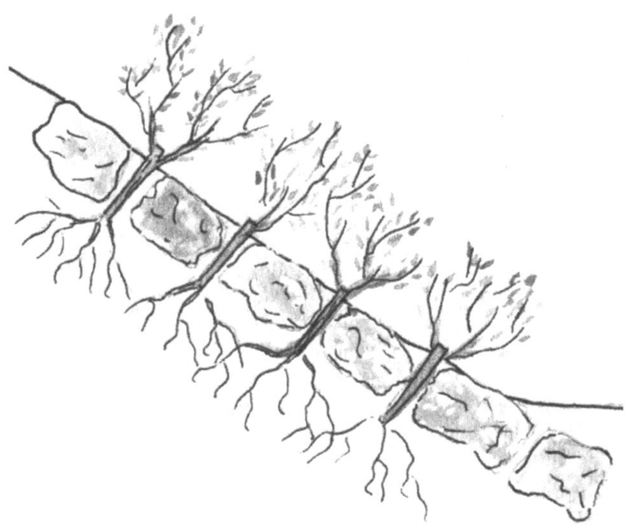		

Table 7-27. Silt fences

<table>
<tr><th>Application</th><th>Advantages</th><th>Disadvantages</th><th>Effectiveness</th></tr>
<tr><td>On disturbed soils such as following a wildfire</td><td>Can be used across a wide range of slope inclinations, covering different slope lengths:

For slope inclination 1:2, the max slope length covered is 15 m, while slopes gentler than 1:5 can be up to 60 m long when covered by a silt fence</td><td>Not effective across drainage ways, gullies, ditches or other areas of concentrated water flow</td><td>Temporary measure that provides barrier to catch the sediment and the runoff from small areas</td></tr>
<tr><th>Material</th><th colspan="3">Diagram</th></tr>
<tr><td>Fence posts (at least 0.90 m long, of hardwood with minimum diameter of 0.08 m if wooden, or a standard T profile if metallic), wire, geotextile fabric

Should be installed on the contour of the slope</td><td colspan="3">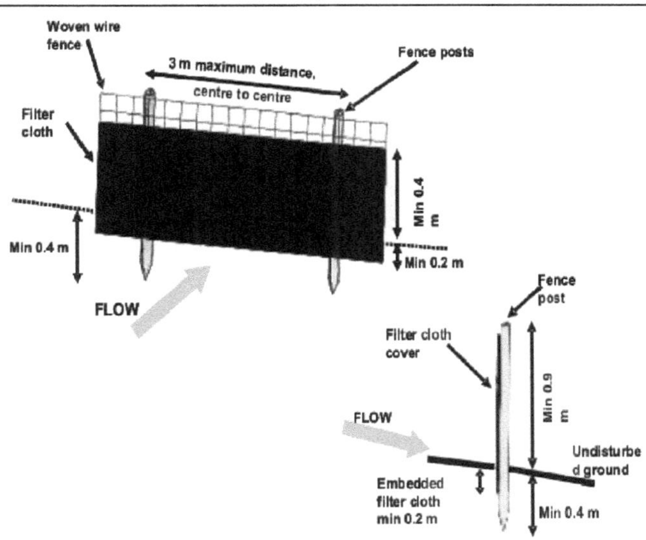
</td></tr>
</table>

Table 7-28. Slope drainage using phreatophytes

<table>
<tr><th>Application</th><th>Advantages</th><th>Disadvantages</th><th>Effectiveness</th></tr>
<tr><td>Wet areas

Suitable in areas of high summer rainfall

In combination with other bioengineering systems</td><td>Simple and economical method in large wet areas

Pumping plants can be used to drain deeper layers in the ground</td><td>Effective only after the plants have rooted</td><td>The plants draw most of the water they need for survival out of the ground

The individual roots work as pumps</td></tr>
<tr><th>Material</th><th colspan="3">Diagram</th></tr>
<tr><td>Plant species with high water consumption - phreatophytes (deep-rooting plants)

An example of water consumption of a poplar tree is given here to illustrate the reduction in moisture content with the distance from the tree (Greenway 1987)</td><td colspan="3">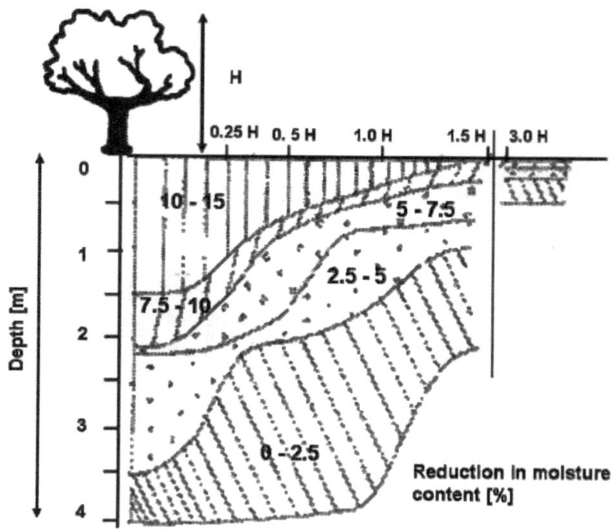
</td></tr>
</table>

Table 7-29. Sodding or turfing

Application	**Advantages**	**Disadvantages**	**Effectiveness**
On slopes with an angle between 0-30° after a medium or high intensity fire Large sandy areas	Perform very well in dry climate Cheap Does not leave permanent patterns on the landscape after removal Increases soil moisture storage >20%	Not effective on steep slopes Not applicable on slopes with rock face	Slowing water movement Provides open channels for water penetration into the deep soil Collecting the sediment, sand and stones moving downwards from the slopes Stopping soil erosion during heavy rainfall Provides both wind breaks to trap seeds and dust and shade and cover for seedlings
Material	**Diagram**		
Hand dug sod slabs: square pieces of 0.40 by 0.40 m are cut out of meadows with more soil (0.08 m thick) **Commercial sod**: the sods are available in strips of 0.3 to 0.4 by 1.5 to 2 m, 0.02 to 0.04 m thick	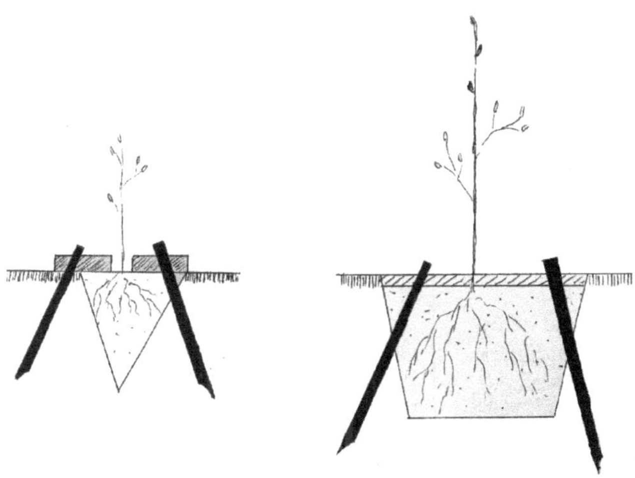		

Table 7-30. Straw bale or wattle check dams

Application	Advantages	Disadvantages	Effectiveness
On gentle slopes after a high or very high intensity fire	Relatively low cost On a slope 0-15° the max drainage between check dams can be up to 4000 m^2 and the maximum slope length up to 60 m. On a slope 15-20° the max drain area between check dams can be up to 2000 m^2 and the maximum slope length up to 30 m.	Not suitable for protection from large storm events or for controlling debris flow in water bodies such as creeks, streams and rivers Not recommended for usage on slopes with inclination greater than 20° Should be very carefully applied avoiding any kind of aggressive treatments	Straw bales are placed in small drainages acting as a dam, collecting upslope sediments and slowing the velocity of water down slope
Material	**Diagram**		
Straw bales or wattles placed in rows with overlapping joints (like a brick wall) Some excavation is necessary to ensure bales butt up tightly against one another forming a good seal Two rows (or walls) of bales are necessary and should be embedded below the ground line at least 0.30 m. The bales and the stakes should be removed once the permanent drainage and stabilization is re-established	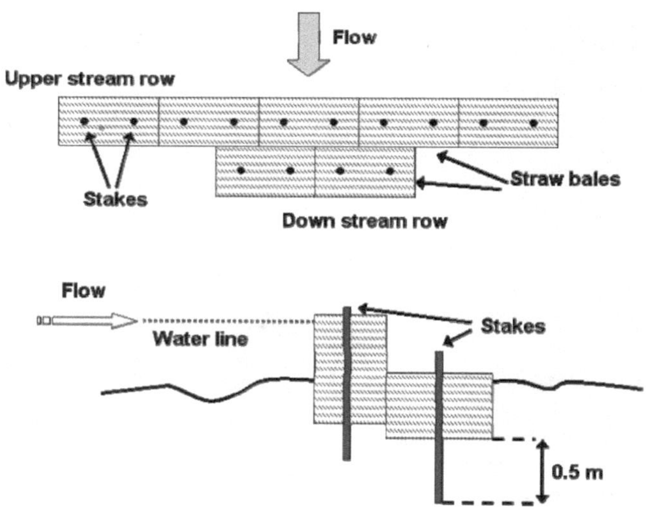		

Table 7-31. Vegetated gabions

Application	**Advantages**	**Disadvantages**	**Effectiveness**
To secure unstable slopes (erosion gullies, banks) To provide drainage through water absorption and transpiration Used in wet areas of fine-grained soil (schistose, clayey, silty substrates)	Fast Simple construction Elastic Can be erected along horizontal lines on wet slopes or along stream channels	Only applicable where gravel and small rocks are available	Gabions form solid protection points There is no danger of water impoundment The plants improve drainage through water absorption and transpiration
Material	**Diagram**		
Wire mesh (0.05m) (right diagram) Steel mesh (left diagram) Coarse gravel Wire for tying Steel pegs Live branches Rooted plants	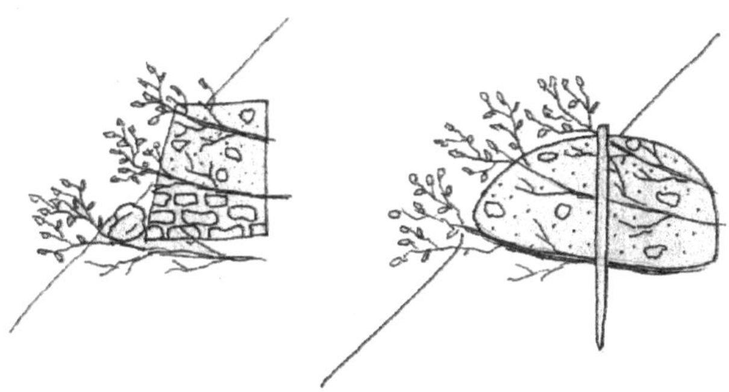		

Table 7-32. Vegetated geogrid

Application	Advantages	Disadvantages	Effectiveness
Similar to branchpacking except that natural or synthetic geotextile materials are wrapped around each soil lift and live branch cuttings are placed between them. For rebuilding very steep eroded streambanks or configuring new banks in stream realignment projects with slopes too steep for normal brushlayering Particularly useful where land has been previously lost and needs to be restored	Efficient minimization of bank erosion Higher initial tolerance of velocity than traditional brushlayering techniques	Systems over 2 m in height and 6 m in length should be subjected to engineering slope stability analysis This technique requires both heavy equipment and intensive manual labour to install	Provide immediate soil reinforcement produce rapid growth, offering overhanging material for aquatic habitat Once the live cuttings become established, their root systems penetrate the grids and the entire system becomes a cohesive mass Improve habitat for aquatic plants and animals Contribute to food web dynamics Enhance aesthetics through the establishment of vegetation
Material	**Diagram**		
Dormant branches from 0.015 to 0.05 m in diameter, long enough to reach the back of the trench to be filled and to extend slightly beyond the surface of the completed slope Geotextile, live stakes and dead stakes, and plants to be installed on top of slope are also necessary	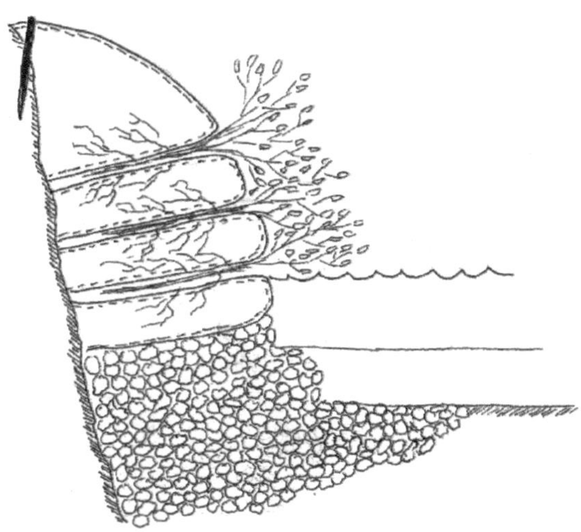		

Table 7-33. Vegetated palisade and pole construction

Application	**Advantages**	**Disadvantages**	**Effectiveness**
In areas of abundant growth (river terraces, forests) Effective method to wall deep and steep V-gullies stair wise with live material Repair of erosion damage in soft fine soils (clay, loess, sand)	Quickly and easily built Immediately effective Exhibits excellent growth	Limited width (6 m) and height (2 to 4 m) Can only be constructed in areas of favorable plant growth	Stabilizes the gully or water channel and causes silting Has an immediate effect as a barrier even before rooting The poles root and pump up water for their growth
Material	**Diagram**		
Pegs or poles from live plants with a diameter of 0.05 m min. (5 to 20 pieces per running meter of construction)	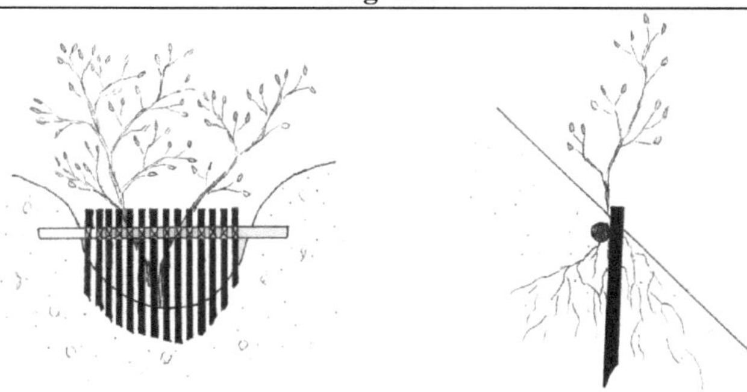		

Table 7-34. Vegetated stone walls and rock piles

Application	**Advantages**	**Disadvantages**	**Effectiveness**
Stabilization of slope parts (toe of the slope) Stabilization of gullies and banks	Possibility of using rubble of mediocre quality and of any size Low cost This construction has flexibility, permeability, and durability Better than non-vegetated stone walls and piles	Possible only during the dormant season of vegetation Wall height is limited	The stone walls and piles with branch layering remain not only permeable, but the plant roots also absorb and transpire a large quantity of water, ensuring drainage, plus the vegetation stabilizes the construction
Material	**Diagram**		
Rocks Slender live branches (2 to 5 per square m) Rooted shrubs (not trees!)			

Table 7-35. Wattle fences

Application	**Advantages**	**Disadvantages**	**Effectiveness**
For the retention of topsoil in minor soil slippages Good in combination with other bio-engineering methods (drainage methods, bank stabilization)	Can be used for mild gully erosion control Can serve as slope drain when wattle fences are arranged with an angle Provide a possible way of stopping the moving materials on slope With the interwoven branches, solid steps can be built into the slope	Unable to stop deep soil movement Large quantity of plant materials Only long flexible branches can be used The branches lie partially on the surface and do not root at all Water can easily penetrate into the soil and cause slippage. The pegs easily broken by a rockfall. High labour and material costs More readily available measures exist for slope stabilisation	Continuously laid packed bundles of plant material intercept surface water runoff and divert it laterally before it creates erosion problems The wattles help trapping sediment to protect downslope areas from material falls or erosion

<table>
<tr><th>Material</th><th></th><th>Diagram</th><th></th></tr>
<tr><td>Flexible branches with few side branches (1.20 m) preferably (shrubby willows)
Wooden or steel pegs 1 m long.

Combination of live and dead pegs less than 1 m long
Plants that root easily from cuttings should be used</td><td colspan="3">
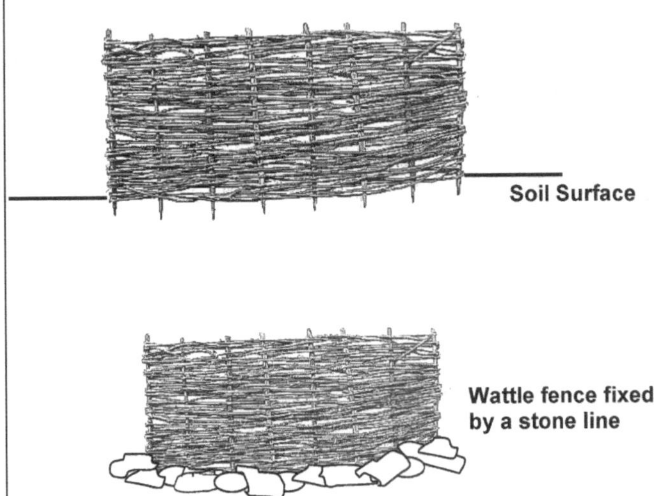

</td></tr>
</table>

3. ECO-ENGINEERING TECHNIQUES

3.1 Management strategies for limiting erosion

Techniques have been developed to maintain or to minimize erosion rates to levels below the soil generation rates. Their objectives are mainly to avoid or to compensate erosion losses and the maintenance of sustainable soil productivity and soil ecological functions. It is a theme in which the use of ground bio- and eco-engineering techniques is very concomitant and difficult to differentiate. Generally, management practices are focused on these main tasks (Schiechtl 1980, Coppin and Richards 1990, Gray and Sotir 1996):

- Increasing or maintenance of the vegetation cover
- Improving the soil hydrology
- Increasing the soil structural stability
- Increasing the surface roughness
- Physically slowing down of erosion dynamics
- Compatibility with traditional management systems

The role of vegetation in erosion control can be summarized as:

Protective role of vegetation	• Interception of the rainfall • Restraint • Retardation of runoff • Infiltration
Most effective vegetation for erosion control	• Herbaceous plants • Grasses and shrubs, possibly with dense near surface root mat and good surface cover and foliage

The principles, when designing a prevention and control system, are based on the basic knowledge of the biophysical characteristics of the intervention area, and the common sense and their application in combination with one or more particular erosion control measures. In many cases, ground bio- and eco-engineering methods can be complementarily applied to increase the effectiveness of the actions realized.

General principles are:

- Extensive grading and earthwork in erosion prone areas or slopes should be avoided
- Increased runoff should be handled with installed hydraulic conveyance facilities

- Runoff velocities should be kept as low as possible
- Soil moisture should be maintained as much as possible
- Interceptor drains and berms should be constructed to divert the runoff away from steep and bare slopes
- Native vegetation on the site should be saved and protected where possible
- If the vegetation needs to be cleared, this should be done in small workable increments, keeping the duration of exposure as short as possible
- Cleared areas should be protected with mulches and temporary fast growing herbaceous covers
- Sediment basins should be constructed in order to prevent eroded soil or sediment from leaving the site
- Erosion control measures should be applied as soon as possible
- The erosion control measures should be surveyed and maintained regularly

In this sense, the most used management practices to prevent or reduce erosion are:

Crop Management

- Crop rotation, choosing a crop sequence that maintains the residue cover (e.g. double-cropping or use of winter cover crops)
- High density planting to create a thick cover for soil protection
- Multiple cropping, by combination of crops with different morphological structures and heights
- Mulching, by addition of crop residues, straw, "green amendments", etc. to the soil surface
- Using conservation tillage, which basically is the tillage/planting system that leaves at least 30% of the field surface covered with crop residue after planting, has been completed.
- Using contour tillage, contour ploughing and wind breaks
- Avoiding overgrazing and the over-use of crop lands
- Selecting crops that produce large amounts of residue (corn grain/*Zea mays* L., sorghum/*Sorghum vulgare* (L.) etc) and/or a high degree of soil cover per kilogram of residue (e.g. wheat/*Triticum aestivum* L.)

Vegetation Management

- Revegetation by planting adequate native species of shrubs and grasses
- Reforestation

- Using agro-forestry techniques
- Planting shrubs or native vegetation to grow along the river banks instead of ploughing and planting crops right up to the water's edge
- Applying bioengineering techniques (Tables 7-1–7-35)
- Leaving unploughed grass strips between ploughed lands
- Planting appropriate vegetation in areas where erosion is most concentrated (see Chapter 6)

Soil management

- Application of organic amendments
- Using soil stabilizers
- Preventing soil compaction
- Preparing adequately the soil-hole for planting (Table 7-1)
- Applying minimum or no tillage practices
- Using crops that provide long-lasting residues (i.e. crops with a high carbon-to-nitrogen ratio, e.g. wheat).
- Surface soil mulching (Table 7-25)

Mechanical methods

- Contouring structures (Tables 7-8 and 7-9)
- Terracing (bench terraces, mini-terraces, etc; wattle fences, logs, etc) (Table 7-35)
- Stabilisation structures (e.g. retaining walls) (Table 7-34)
- Ditches, berms (Table 7-11)

3.2 Eco-engineering techniques against rockfall

An excellent alternative for technical protective constructions against rockfall can be provided by different types of forest stands, given the urgency of the protection needed and the site conditions that determine forest stand development. The management of protection forests is to a large degree a trade-off between optimizing the protective effect and assuring forest stand stability at present and over the long-term (Motta and Haudemand 2000). Since stand stability is mostly at risk in over-mature stands that lack sufficient regeneration, management interventions in rockfall protection forests often aim at thinning or creating gaps to allow more light into the forest stand. To increase terrain roughness, a common recommendation in rockfall protection forest management is to leave the trunks of cut trees lying on the slope, preferably diagonally to the slope direction, to create obstacles (Mössmer et al. 1994, Dorren et al. 2005, Frehner et al. 2005). These diagonally positioned logs prevent the development of rock accumulations and allow continued rock transport in a controlled manner. Experience in Austria is that larger *Picea abies* trees (DBH > 50 cm) can act as effective rockfall barriers for approximately 10 years (Dorren et al. 2005). Additionally, high tree stumps (e.g. > 1.3 m) have been noted to further reduce residual rockfall hazard on a site (cf. Dorren et al. 2005; Frehner et al. 2005).

To give a guideline for the different options for using eco-engineering techniques against rockfall, the optimal forest cover type for each characteristic rockfall zone is discussed. These are 1) the rockfall source area, 2) the transport zone and 3) the rockfall accumulation or deposit area (Figure 7.1). The optimal forest cover type will be discussed in terms of structure and tree species.

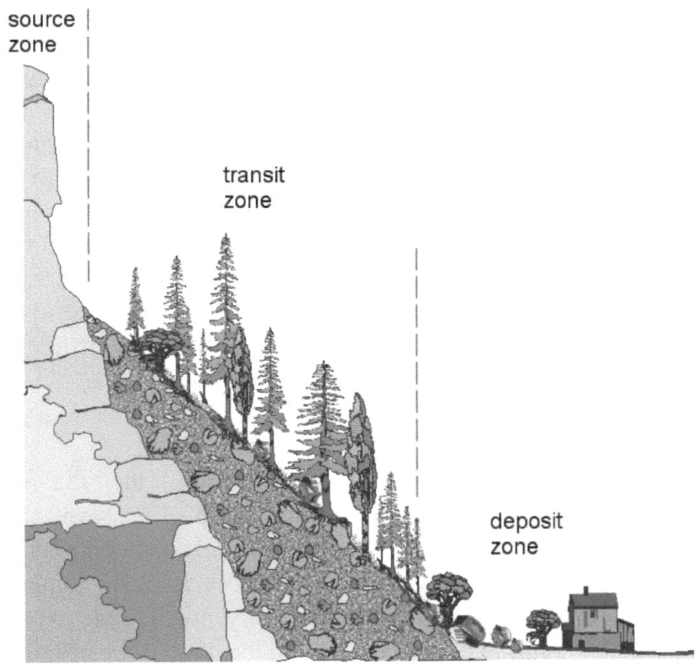

Figure 7-1. Characteristic zones on an active rockfall slope.

Source area

Rockfall source areas are generally characterized by steep cliff faces that show unfavorable combinations of the exposition of the slope face with the dip and strike of the bedding planes and the most prominent joint sets. Root actions of large trees can increase the production of individual falling rocks.

Therefore, large trees growing on top or in vertical cliff faces should be removed. In case of a stepped terrain, where vertical cliff faces and more horizontal areas occur on top of each other, trees do not necessarily promote rockfall activity by their roots. Moreover, they can reduce the initial velocity and jump height of falling rocks. In such cases they should be examined to ascertain whether they do have a rockfall promoting effect, before removing them. We do not recommend any specific forest management actions other than the removal of trees if necessary. If cut tree stems can be put in a stable position, diagonal to the slope direction, additional rockfall barriers can be created.

Transport zone

The rockfall transport zone lies in between the rockfall source area and the deposition area. In this zone the rockfall velocities as well as the jump heights are maximal. Consequently, the objective of rockfall protection in this zone is to reduce both of them or, in an optimal case, to stop the falling rock. The first guidelines for achieving the latter using a forest stand were published by Wasser and Frehner (1996). They recommended a forest stand consisting of more than 400 trees per ha with diameters larger than 40 cm. In the European Alps, such a forest, however, consists mostly only in stands with a regular structure. Such stand structures are not stable in the long-term and therefore cannot provide sustainable mitigation. Irregular forest stands consisting of trees of various ages and diameters and preferably mixed species are much more stable and provide better protection in the long-term. The question is then, what type of stand structure (density of trees, species, spatial distribution of diameter) is needed? The answer to this question depends on the average size of the falling rocks and the slope angle. These two factors determine the energy that has to be dissipated.

Rockfall experiments on forested slopes showed that the number of impacts against trees is more important than the efficacy of the impact expressed in the amount of dissipated energy (e.g. Berger et al. 2002; Dorren et al. 2005). Therefore, a large number of trees is more important than having only thick tree stems. Again, diagonally positioned tree stems can have the same effect as standing trees and reduce the energy of falling rocks. The larger the tree the more energy can be dissipated. This resulted from a large number of real size rockfall experiments on both non-forested slopes and forested slopes with different forest types. Experimental slopes had a slope angle between 38° and 42°, which is typical for forest covered rockfall talus slopes and rockfall transport zones. The guidelines given here are certainly valid for less steeper slopes. For steeper slopes, a greater number of trees is needed, but this is often difficult as site conditions do not allow that. The rock size used in the experiments varied between diameters of 25 cm to

125 cm. Tables 7-36 and 7-37 are presented to assist in the design of optimal protection forests against rockfall. The initial data needed to design the protection forest is the average energy of the falling rock, as shown in Table 7-36; this can be calculated from the average diameter of the falling rock. This allows calculation of the mass (assuming a rock density of 2800 kg/m^3) and the energy, given a certain velocity related to the initial fall height. Subsequently, Table 7-37 provides information about the amount of energy that can be dissipated during a single frontal impact on different types of trees as derived from Dorren and Berger (2006). Frontal impacts on trees are the most effective and scratch impacts (impacts on the side of the tree stem) are least or almost not effective in terms of energy dissipation).

Scratch impacts, however, do cause lateral deviations in the rockfall trajectory, as seen from the slope direction, causing the rock to travel a longer distance in the forest. As a result the chance of the rock impacting a tree increases. On our study sites, the forest cover reduced the rockfall velocity by 20% and the jump heights by 60%. However, it also results in lateral deviation and therefore a wider runout zone. For safety reasons, we take into account a runout zone as shown in Figure 7.2, which means a lateral deviation of 10° from the straight downslope direction to both sides.

Analysis of the results of the real size rockfall experiments in a mixed forest covering a slope with a mean slope gradient of 38° showed that the average distance between two tree impacts was 31.7 m. This is the first important condition to assess the required structure of a rockfall protection forest stand.

Next a procedure is needed that translates the spatial distribution of the tree diameters and the number of trees per hectare into the probable distance between two subsequent impacts against trees. We developed a simple method, adapted from the Mean Tree Free Distance concept of Gsteiger (1993), which assumes that a certain forest structure can be expressed in a virtual sequence of rockfall protective tree nets (curtains) consisting of a row of trees perpendicular to the direction of the slope, as shown in Figure 7.3. The distance between two trees in one virtual row is 90% of the diameter of the average falling rock (represented by a sphere with the equivalent volume). By using the average tree diameter, the existing forest structure can be expressed in a number of virtual tree nets, which is equal to the number of probable impacts.

By knowing the minimal distance between rock impacts and the number of impacts needed to stop a falling rock, the total number of trees and their average diameter can be calculated using the above principle. If, in addition, the slope length is known, the number of trees in the transport area can be calculated, using the 20° angle area shown in Figure 7.2. This number of

Table 7-36. Average size of the falling rock and its energies (in grey: values that are not realistic)

Diameter (cm) **Volume (m^3)** **Mass (kg)**	25 0.008 23			50 0.065 183			100 0.524 1466			125 1.023 2863		
Energy (kJ)	**Velocity ($m\ s^{-1}$)**	**Velocity ($km\ h^{-1}$)**	**Fall height (m)**	**Velocity ($m\ s^{-1}$)**	**Velocity ($km\ h^{-1}$)**	**Fall height (m)**	**Velocity ($m\ s^{-1}$)**	**Velocity ($km\ h^{-1}$)**	**Fall height (m)**	**Velocity ($m\ s^{-1}$)**	**Velocity ($km\ h^{-1}$)**	**Fall height (m)**
10	30	106	44	10	38	6	4	13	1	3	10	0.4
100	93	336	445	33	119	56	12	42	7	8	30	4
500	209	752	2225	74	266	278	26	94	35	19	67	18
1000	295	1064	4450	104	376	556	37	133	70	26	95	36
1500	362	1303	6675	128	461	834	45	163	104	32	117	53

Table 7-37. The effective tree diameter [cm] for a total energy dissipation in a single impact

Energy (kJ)	***Picea abies***	***Abies alba***	***Acer pseudoplatanus***	***Fagus sylvatica***
10	12	11	11	9
100	31	30	29	24
500	63	60	57	48
1000	85	81	77	65
1500	101	97	92	78

Figure 7-2. Lateral deviation of the falling rock on forested slopes results in wide runout zones. An angle of 20° has to be taken into account as shown in the figure.

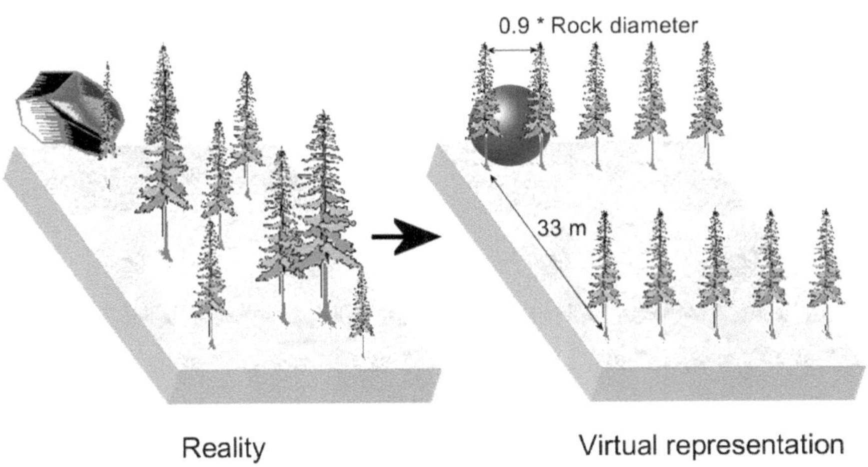

Figure 7-3. Explanation of the principle for expressing a real forest structure in a sequence of virtual rockfall protective tree nets (curtains).

trees can then be expressed in the number of trees per hectare in the transport area. Combining this number with the average diameter provides the volume. This above described method forms the basis for Tables 7-38 and 7-39. These tables provide guidelines for the number of trees per hectare and their minimal average effective diameter for a given slope length and for a given rock diameter. These data are given both for spruce and for beech on a slope of 40° or less. The minimal slope length in the tables is 100 m as the data analysis showed that for a slope length of 50 m the required forest structure (expressed in stem density and diameters) to stop a falling rock with a diameter of 1 m and an energy between 500-1000 kJ is not realistic. This is shown in Figure 7.4. Similar analyses can be performed online, using the free and publicly available tool at www.rockfor.net.

Deposition area

In the rockfall deposition area, the same guidelines can be used as in the transport zone, but the diameters can be smaller. It is more important that a lot of trees occupy this zone, e.g. coppice stands, and that the surface is as rough as possible (e.g. deposited rocks, cut tree stems). Therefore, regeneration has to be promoted, preferably fast growing species combined with strong rockfall resisting trees such as beech and sycamore. A dense forest stand with tree diameters of 10 cm could already be effective here.

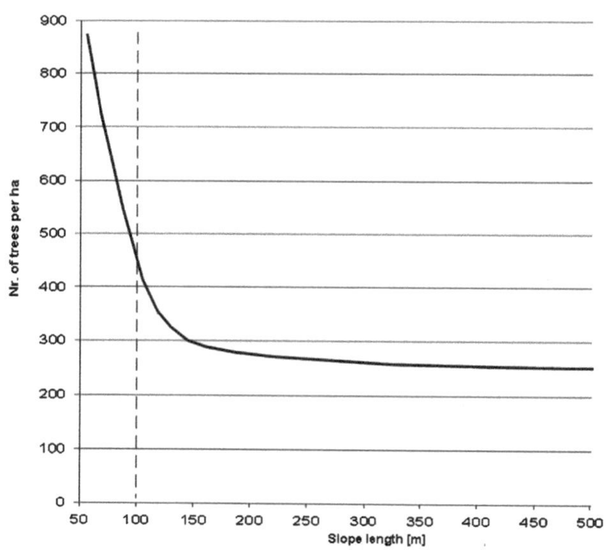

Figure 7-4. Slope length versus the number of trees per hectare (average diameter of 35 cm) needed to stop a falling rock with a diameter of 1 m and an energy between 500-1000 kJ. The figure shows that a minimum slope length of approximately 100 m is required for a realistic, sustainable protection provided by forests.

Table 7-38. Proposed parameters for a protection forest consisting of spruce trees (in grey: values that are not realistic)

Spruce 100m slope length	Rock diameter [cm]	25			50			100			125		
Energy [kJ]	**Energy to be dissipated by impact[kJ]**	**Min Average tree Diameter [cm]**	**Min Stem density [stem/ha]**	**Min Basal area [m²/ha]**	**Min Average tree Diameter [cm]**	**Min Stem density [stem/ha]**	**Min Basal area [m²/ha]**	**Min Average tree Diameter [cm]**	**Min Stem density [stem/ha]**	**Min Basal area [m²/ha]**	**Min Average tree Diameter [cm]**	**Min Stem density [stem/ha]**	**Min Basal area [m²/ha]**
10	3	10	1846	14	10	1091	9	10	600	5	10	490	4
100	30	15	1600	28	15	1000	18	15	571	10	15	471	8
500	150	35	1043	100	35	750	72	35	480	46	35	407	39
1000	300	50	828	162	50	632	124	50	429	84	50	369	72
1500	450	60	727	206	60	571	162	60	400	113	60	348	98
Spruce 250m slope length	**Rock diameter [cm]**	**25**			**50**			**100**			**125**		
Energy [kJ]	**Energy to be dissipated by impact[kJ]**	**Min Average tree Diameter [cm]**	**Min Stem density [stem/ha]**	**Min Basal area [m²/ha]**	**Min Average tree Diameter [cm]**	**Min Stem density [stem/ha]**	**Min Basal area [m²/ha]**	**Min Average tree Diameter [cm]**	**Min Stem density [stem/ha]**	**Min Basal area [m²/ha]**	**Min Average tree Diameter [cm]**	**Min Stem density [stem/ha]**	**Min Basal area [m²/ha]**
10	3	10	1063	8	10	628	5	10	346	3	10	282	2
100	30	15	922	16	15	576	10	15	329	6	15	271	5
500	150	35	601	58	35	432	42	35	276	27	35	234	23
1000	300	50	477	94	50	364	71	50	247	48	50	213	42
1500	450	60	419	118	60	329	93	60	230	65	60	200	57

Table 7-38. (Continued)

Spruce 500-1000m slope length	Rock diameter [cm]	25			50			100			125		
Energy [kJ]	Energy to be dissipated by impact[kJ]	Min Average tree Diameter [cm]`	Min Stem density [stem/ha]	Min Basal area [m^2/ha]	Min Average tree Diameter [cm]	Min Stem density [stem/ha]	Min Basal area [m^2/ha]	Min Average tree Diameter [cm]	Min Stem density [stem/ha]	Min Basal area [m^2/ha]	Min Average tree Diameter [cm]	Min Stem density [stem/ha]	Min Basal area [m^2/ha]
10	3	10	1004	8	10	593	5	10	326	3	10	266	2
100	30	15	870	15	15	544	10	15	311	5	15	256	5
500	150	35	568	55	35	408	39	35	261	25	35	221	21
1000	300	50	450	88	50	344	67	50	233	46	50	201	39
1500	450	60	396	112	60	311	88	60	218	62	60	189	53

Table 7-39. Proposed parameters for a protection forest consisting of beech trees (in grey: values that are not realistic)

Beech 100m slope length	Rock diameter [cm]	25			50			100			125		
Energy [kJ]	Energy to be dissipated by impact[kJ]	Min Average tree Diameter [cm]	Min Stem density [stem/ha]	Min Basal area [m²/ha]	Min Average tree Diameter [cm]	Min Stem density [stem/ha]	Min Basal area [m²/ha]	Min Average tree Diameter [cm]	Min Stem density [stem/ha]	Min Basal area [m²/ha]	Min Average tree Diameter [cm]	Min Stem density [stem/ha]	
10	3	5	2182	4	5	1200	2	5	632	1	5	511	
100	30	15	1600	28	15	1000	18	15	571	10	15	471	
500	150	25	1263	62	25	857	42	25	522	26	25	436	
1000	300	35	1043	100	35	750	72	35	480	46	35	407	
1500	450	45	889	141	45	667	106	45	444	71	45	381	
Beech 250m slope length	**Rock diameter [cm]**	**25**			**50**			**100**			**125**		
Energy [kJ]	Energy to be dissipated by impact[kJ]	Min Average tree Diameter [cm]	Min Stem density [stem/ha]	Min Basal area [m²/ha]	Min Average tree Diameter [cm]	Min Stem density [stem/ha]	Min Basal area [m²/ha]	Min Average tree Diameter [cm]	Min Stem density [stem/ha]	Min Basal area [m²/ha]	Min Average tree Diameter [cm]	Min Stem density [stem/ha]	Min Basal area [m²/ha]
10	3	5	1257	2	5	691	1	5	364	1	5	294	1
100	30	15	922	16	15	576	10	15	329	6	15	271	5
500	150	25	728	36	25	494	24	25	301	15	25	251	12
1000	300	35	601	58	35	432	42	35	276	27	35	234	23
1500	450	45	512	81	45	384	61	45	256	41	45	219	35

Table 7-39. (Continued)

Beech 500-1000m slope length	Rock diameter [cm]	25			50			100			125		
Energy [kJ]	**Energy to be dissipated by impact[kJ]**	**Min Average tree Diameter [cm]**	**Min Stem density [stem/ha]**	**Min Basal area [m²/ha]**	**Min Average tree Diameter [cm]**	**Min Stem density [stem/ha]**	**Min Basal area [m²/ha]**	**Min Average tree Diameter [cm]**	**Min Stem density [stem/ha]**	**Min Basal area [m²/ha]**	**Min Average tree Diameter [cm]**	**Min Stem density [stem/ha]**	**Min Basal area [m²/ha]**
10	3	5	1187	2	5	653	1	5	344	1	5	278	
100	30	15	870	15	15	544	10	15	311	5	15	256	5
500	150	25	687	34	25	466	23	25	284	14	25	237	12
1000	300	35	568	55	35	408	39	35	261	25	35	221	21
1500	450	45	484	77	45	363	58	45	242	38	45	207	33

3.3 Management strategies to protect against windthrow

The contribution of a forest stand to the reduction of erosion, rockfall and landslide risks on a slope will change over time. It is the role of forest managers to understand how and why these changes take place, and to devise silvicultural scenarios accordingly. These scenarios should describe the methodology by which a forest stand will be tended, harvested and replaced; a process often categorized into 'silvicultural systems', according to the chosen reproduction method (Daniel et al. 1979).

Forest managers have to ensure that the silvicultural system they choose meets their management objectives but remains within given environmental and operational constraints. Silvicultural textbooks (e.g. Daniel et al. 1979; Smith et al. 1997; Nyland 2002) provide detailed explanations on how different silvicultural systems can be applied to different situations. In the following sections we highlight how this choice will affect the distribution of the risk of significant wind damage over time. As part of these silvicultural systems, several forest operations are used to maintain, harvest and regenerate stands. We therefore describe operational strategies that may be used to minimise risk to forest stands in wind exposed situations.

3.3.1 Silvicultural systems and wind risk

High forest

1. Even-aged stands

Even-aged stands are those which are regenerated at once or, in the case of naturally regenerated stands, during a short period corresponding to less than 1/5th of the full rotation period.

2. Clear-felling system

This system is characterized by the harvesting of all trees in the stand at the end of the rotation period. As risk increases with tree height and age (see Chapter 5), the risk of wind damage to an old even-aged stand is likely to be high (Figure 7.5). The increase of risk with time is a factor to take into consideration when afforesting an unstable slope with young trees. Under such a system, the role of the forest manager is to choose a suitable rotation age for regenerating the stand. One important drawback of this system is that there is a period between the clear felling and the reforestation when the site has completely lost its tree cover. The risk this creates for slope stability may be mitigated by using variants of the same system, i.e. by designing the spatial distribution of felling coupes so that the slope keeps some tree cover. For example, the forest manager may choose to harvest the trees in alternate strips.

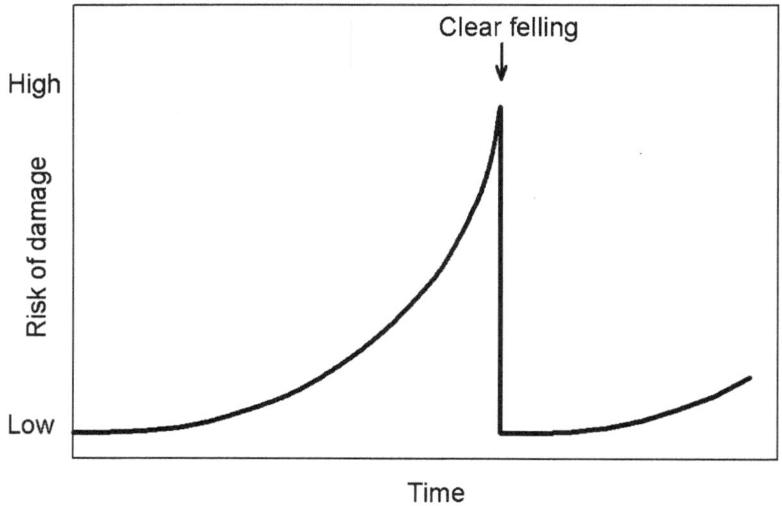

Figure 7-5. Changing risk of wind damage to a stand managed with the Clear-felling System.

3. Seed-tree system

This system is similar to the clear-felling system, except that a small number of trees are left standing in order to provide a seed source that will help regenerate the site naturally. This will result in increased wind loading to the seed trees and, as they are not adapted to the new conditions, wind damage will be common in the first few years after the harvest (Figure 7.6).

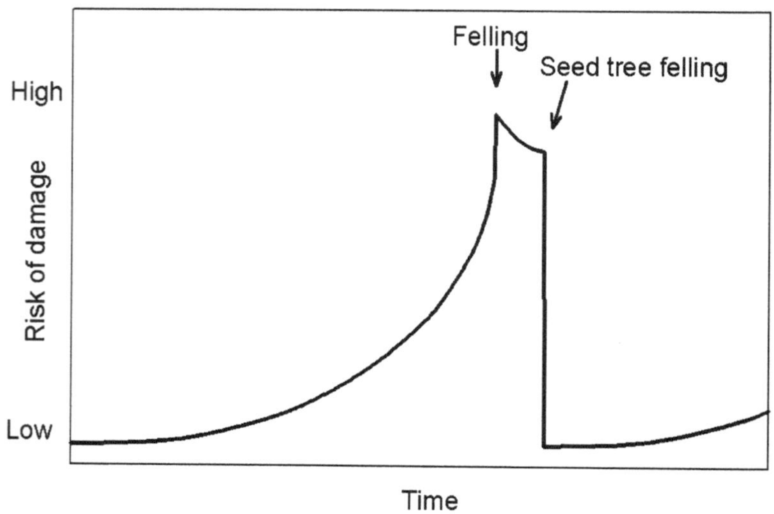

Figure 7-6. Changing risk of wind damage to a stand managed using the Seed-tree System.

However, if no or limited damage occurs, a site managed under this system will always contain some trees – although the amount of slope protection provided by the seed trees is likely to be limited.

4. Shelterwood system

Under this system, the site is naturally regenerated through a series of (generally two or three) partial cuts which aim to provide a seed source and the right conditions in the understory for natural regeneration to be established. The final harvest is conducted only when there is sufficient natural regeneration under the mature crop and therefore the site will always maintain a tree cover. The risk of wind damage (Figure 7.7) after the partial cuts will increase, but the gradual opening of the stand will reduce the likelihood of a catastrophic event (Gardiner et al. 2005). As for other systems, the manager may decide to distribute the coupes into different spatial patterns. Commonly used methods include uniform, group and strip shelterwood. In the first case individual stems are harvested across the site, in the second case stems are harvested in small groups, and in the last case they are harvested in strips. The latter may be applied to slopes where the risk of wind damage is considered high. A good strategy is to work in successive strips going towards the main wind direction but the applicability of this on a slope will depend on its orientation. Care must be taken when performing the harvesting operations in order to avoid damaging the regeneration.

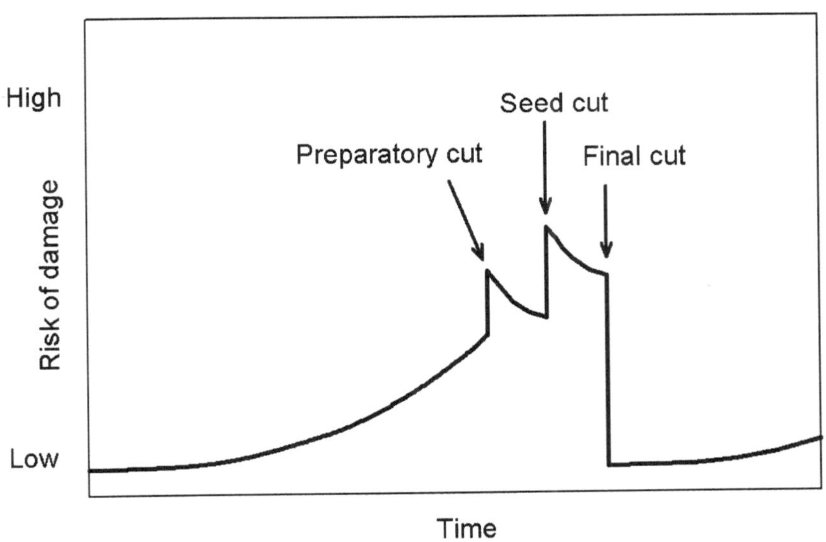

Figure 7-7. Changing risk of wind damage to a stand managed using the Shelterwood System.

Uneven-aged stands

An uneven-aged stand is an area containing trees at different stages of development. Silviculturists often perpetuate a stand structure where the distribution of trees in different age classes leads to a distribution in a 'reverse-j', i.e. the number of stems decreases with increasing age class. This is the 'selection system'.

5. Selection system

The main characteristics of this system are that 1) it perpetuates the uneven-aged structure of the stand and 2) the regeneration is always protected by the presence of older trees. It will therefore maintain a good tree cover through time which is an advantage for the protection of very sensitive slopes. Generally, a cycle of three or four harvesting operations will be planned during a full rotation. There are several variants that exist within this system, depending on how these operations are conducted. For example, these can involve the removal of individual stems (single-tree selection) or groups of stems (group-selection). The risk of wind damage to trees will be heightened after each intervention in the forest, but because of the continuous presence of young trees in the stand the risk of an event completely destroying the tree cover is relatively low (Figure 7.8).

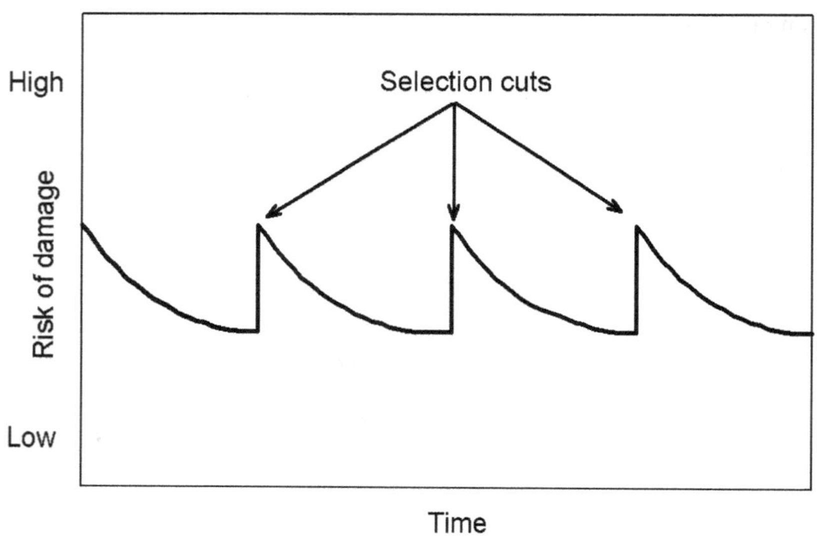

Figure 7-8. Changing risk over time of significant wind damage in a stand managed using the Selection System.

6. Coppice forest

Coppice forests are those which are regenerated through the vegetative sprouting of buds following the harvesting of the stem. Sprouts can originate from the stump or the root system of the tree. They are particularly vigorous in some species so that the coppice regeneration method is often used when the site is dedicated to the production of biomass. The harvesting of the trees is normally conducted using a clear-felling reproduction method. The system is interesting from the point of view of slope stabilisation because even though the aerial parts of the trees are removed periodically the site will benefit from the continuous presence of a well-developed rooting network. The rotation period is usually short (it can be as little as two to five years). It therefore involves frequent operations which might not be suitable for very sensitive or steep slopes.

3.3.2 Operational strategies

1. Thinning

Forest stands on steep slopes that have high topographical exposure and a high risk of wind damage should be managed carefully to avoid any increase in risk. In particular, stand thinning immediately increases the risk of tree overturning and stem breakage and should be practiced with care in wind exposed stands. The magnitude of this increase in windthrow risk depends on how and when the stand is thinned (Hibberd 1991). The larger the gaps that are created, the greater the increase in risk, so heavy thinning should be avoided on vulnerable sites. Thinning at an early age, i.e. 'precommercial thinning' leads to only slightly increased risk, while thinning more mature stands with a high canopy will make the trees immediately vulnerable to wind damage. However, stands that survive this increased risk will restabilise themselves over subsequent years, with risk commonly returning to pre-thinning levels within five to ten years of thinning, depending on species, yield class and age. An option that avoids a sudden increase in wind risk, is to plant 'self-thinning' mixtures of fast and slower growing tree species. Over time, the faster growing trees shade out the slower growing trees which eventually die out. For example, in the UK, plantations of self thinning mixtures of Scots pine and Sitka spruce have been successful in producing a final crop of well spaced Sitka spruce.

2. Felling

When felling vulnerable stands of trees on slopes, losses to windthrow may be minimised by felling stands with the highest windthrow risk before more stable or more sheltered stands. In addition, it is possible to plan felling operations to avoid or minimise exposing other vulnerable stands, by commencing felling at the downwind end relative to the prevailing wind. If selecting stands of older trees to be retained past their expected felling age, it is advisable to avoid stands on exposed sites, wet soils, and those that are immediately downwind of planned fellings (Quine et al. 1995).

Trees are particularly vulnerable to windthrow where the roots have restricted downward development. Roots compensate to some extent with adaptive growth and production of wider root plates, but shallow rooted trees remain less stable than deep rooted trees (Ray and Nicoll 1998). Early felling of stands may be necessary on soils where rooting depth is limited by a high water-table, induration, strong iron pan or shallow bedrock.

3. Stand edges

The wind loading is higher on trees close to the forest edge than on trees inside the forest. If trees have grown up at a forest edge, they will have adapted to their wind environment and be no more vulnerable than interior trees. The existence of windfirm edges is crucial to successful coupe design in moderate to high wind risk locations, such as on exposed slopes. Edges become windfirm because trees exposed to wind develop buttresses, stronger roots, wider root systems, and greater stem taper (Nicoll and Ray 1996; Cucchi et al. 2004). These stable edges also reduce the penetration of strong winds into the stand. Problems occur when a new edge is created, for example through clear felling or road construction. The newly exposed trees are much more vulnerable to being windthrown, even with relatively low wind speeds, because they are not adapted to their new wind environment (Quine and Gardiner 1992).

Topping (removing up to a third of the top of the crown) or high pruning (removing a third of the lower crown) the edge trees can significantly reduce the risk of wind damage (Hunt and Gardiner 2002). Alternatively, severance cuts can be made a few years ahead of clear felling or road building to precondition the remaining trees to their environments.

Evidence suggests that trees about 4-5 tree heights back from the forest edge are the most vulnerable. This appears to be due to flow distortion of wind at the forest edge and the time it takes for damaging gusts to develop. Modifying the shape of the forest edge at establishment or during management operations in order to create tapered edges (by planting or favouring slower growing species at the edge) or having graduated tree density at the edge can have stability benefits (Figure 7.9). This is because the flow distortion at the edge is minimised.

Tapered edges should be at least ½ tree heights wide and any manipulation to the edge of the stand should bear in mind that remaining trees will take time to adjust to their new environment (Gardiner and Stacey 1996). The shape of the edge also influences the risk of damage, and it is important when designing forests in exposed situations to avoid creating concave stand edges that accentuate the topographic funnelling of the wind.

Although wind loading on edge trees is greater than trees within the stand, growth of the stem and root system more than compensates for this. Severance cuts are designed to prepare and condition the forest for a future harvesting operation (Quine et al. 1995). These can be used to create a wind-firm edge in an exposed upland forest by exposing young trees to increased wind loading so that they adapt their growth in subsequent years.

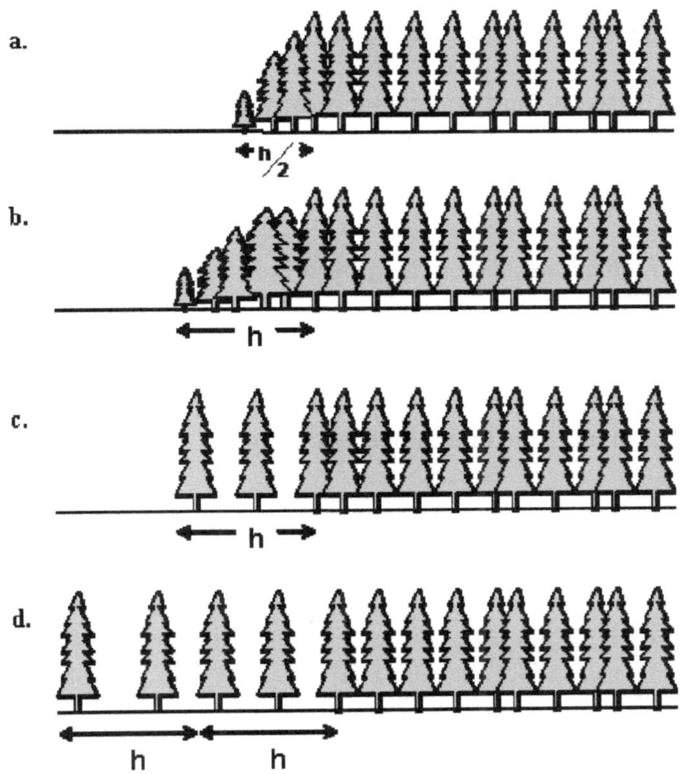

Figure 7-9. Recommended alternative designs of forest edges to improve stand stability by reducing air-flow distortion (adapted from Gardiner and Stacey 1996). a. and b. tapered edges that are ½ and 1 tree heights (h) wide, c. and d. graduated tree density for 1 and 2 tree heights.

Severance cuts should be made during a period when the trees are at low risk, and they should be as wide as the height the trees will be when they are expected to form a new edge. Severance cuts may be combined with respacing to create a graduated density edge.

4. COMBINING GROUND BIO- AND ECO- ENGINEERING TECHNIQUES

Ground bio- and eco-engineering techniques can be combined depending on the particular problem and type of vegetated slope. An example may be the restoration of degraded woodlands due to forest fires, intensive farming or forestry activities and over-grazing. Three groups of restoration activities are required: 1) erosion control works, 2) waterflow control works and 3) vegetation recovery by artificial reforestations and by natural regeneration.

4.1 Erosion control works

These are usually built up on a temporary basis over 5-10 years to stop sheet erosion caused by heavy rain. The ground bioengineering techniques suitable for erosion control works are:

- *Matchsticks* - made of branches, bracelets, thin boles, and the remnants of clear fellings, stacked on the ground in horizontal lines, (on the uphill side of the stumps), in areas of moderate slopes (10-30°). The distance between these rows is usually 8-15 m depending on relief. Dimensions are, height 0.50-0.75 m, width 1.0-1.5 m and length from 1 to >1000 m (see Table 7-24). These structures collect the sediment, sand and stones stones from moving downslope and stop soil erosion during heavy rainfall.
- *Log erosion barriers* - these structures constitute logs of dead trees, stacked on the ground by poles or tied to tree stumps in horizontal lines, in places where slopes are steep (31-50°). A small trench is built upwards to stop soil moving downwards after rainfall. The distance between each log is varied (8-10 m), height and width equals log diameter (about 0.20 m), and length from 1 to >1000 m (see Table 7-8).

These methods can be combined with eco-engineering techniques such as:

- *Clear felling* - in areas with very steep slopes (more than 50°), the dead (burnt) trees should be cut in to pieces of about 1 m in length and distributed across the slope to form log erosion barriers (see Table 7-8).

- *Ploughing and furrowing* - a heavy machine e.g. a Caterpillar with two ploughs at the rear (nails 1.0 m long at a distance of 2.0 m from each other), ploughs the area once horizontally across the slope at a depth of 0.70-1.0 m between the wood stacks where the slope is moderate 0-30°, leaving a furrow. This technique should be applied only in certain circumstances and aims at preventing further soil erosion and also at improving the soil condition. First, by furrowing the ground, water, soil, and sediment are collected into furrows. Secondly by loosening the soil at the ground surface, rainwater is absorbed and penetrates more easily to deeper layers without eroding the surface. Thirdly, the soil is prepared for plantations. In sleep shallow-rocky soils individual holes for tree/bush planting should be considered.

4.2 Waterflow control works

These works are aimed at controlling waterflow by keeping in place the water and sediments that have escaped the erosion control works. The ground bio-engineering constructions that can be put in place are:

- *Small timbered dams* - these are temporary structures used for 7-10 years. They are wooden structures made with logs from dead trees on 2nd and 3rd degree currents. They are usually constructed in certain places along current beds. They are usually 1 m tall and specially stacked in place.
- *Check dams* - these are permanent constructions made of concrete with heights up to 5.0 m, placed at the lower places of the 1st degree current beds.

4.3 Vegetation recovery by reforestation and natural regeneration

A new management perspective that emphasizes a variety of amenities and commodities is needed for woodlands. Today throughout Europe there is an increasing awareness of the necessity to apply and implement management practices that consider the multiple values in the woodlands on the long-term sustainable basis. The new forest ecosystem should be stable, upgraded, adapted to the climatic and soil conditions, more resistant to fire and insect pests, with a normal potential of fauna and flora. For the re-establishment of a future forest the multiple and social uses of woodlands (e.g., watershed management, wildlife, recreation, hunting, aesthetics, education, etc.), as well as the long-term protection from many dangers (e.g., wildfires, soil erosion, storms, etc.) should be considered.

There are two approaches to this aim: 1) artificial reforestation and 2) to protect the natural regeneration.

1. Artificial reforestation

Artificial reforestation is aimed at initially filling the gaps left from natural regeneration, secondly to re-establish species which have disappeared because of human activity, and thirdly to renew and improve vegetation. Species should be indigenous, whenever possible, and be well adapted to the soil and climatic conditions present. Some exotic species may be used in certain circumstances.

Artificial restoration may use conifer and broadleaved species, evergreen and deciduous species depending on the original woodland or forest. Soil and climatic conditions, space, altitude, exposure, and topography must also be taken into consideration (see Chapter 6).

Conifers can be planted on poorer soils. Some typical species are: *Pinus brutia* (Ten.), *Pinus halepensis* Mill., *Pinus pinea* L., *Pinus nigra* L., *Cupressus sempervirens* L., *Cupressus arizonica* Greene, *Cedrus deodara* (D. Don) G. Don, *Cedrus libani* A.Rich, *Cedrus atlantica* (Endl.) Carrière, *Thuja* sp. L.. Broadleaved species are recommended for planting where better environmental and soil conditions exist. Some typical species are:

- *Quercus aegilops* L., *Quercus pubescens* Willd., *Morus alba* L. on southern exposures with low elevation (where soil moisture is low during the summer in particular).
- *Acer negundo* L., *Acer pseudoplatanus* L., *Quercus ilex* L. on northern exposures (where soil moisture is high during the summer in particular)
- *Quercus frainetto* Ten., *Quercus cerris* L. in higher latitudes
- *Celtis australis* L., *Cercis siliquastrum* L., *Fraxinus ornus* L., *Fraxinus excelsior* L., *Acer campestre* L., *Acer negundo* L., *Robinia pseudacacia* L., *Tillia tomentosa* Moench., *Carpinus orientalis* Mill., *Ulmus* sp. L., in certain places and all over the planted area.

Secondary species such as *Laurus nobilis* L., *Spartium junceum* L., *Rosmarinus officinalis* L., *Nerium oleander* L., *Ligustrum vulgare* L., *Cotoneaster horizontalis* Dcne., *Pyracantha coccinea* M.Roem., *Pyrus malus* L., *Prunus insisititia* L., should be established around recreation sites, fire lanes and forest roads.

2. Natural regeneration

Many species are well adapted to regenerate after wildfire, for example the following conifers; *Pinus brutia* Ten., *Pinus halepensis* Mill., *Pinus pinea* L., *Pinus pinaster* Ait., *Pinus radiata* D.Don, *Cupressus sempervirens* L., *Cupressus arizonica* Greene. Although, in many cases, natural regeneration maybe at risk, because of the possibility that a high percentage of seedlings weaken from drought and pests.

The main understory sprouted shrubs, that adapt to regeneration after wildfires are: *Quercus coccifera* L., *Phillyrea latifolia* L., *Pistacia lentiscus* L., *Pistacia terebinthus* L., *Arbutus unedo* L., *Arbutus andrachne* L., *Paliurus spina-christi* Mill. and *Anthyllis hermanniae* L., (Spanos et al. 2000).

5. CONCLUSION

In this Chapter, we have presented many different ecotechnological solutions including the traditional ground bio-engineering techniques for combating mass movements especially soil erosion, shallow slope instability, rockfall and windthrow. The success of these ecotechnological solutions is very much dependent on local conditions and site-specific factors, therefore it cannot be assumed that each technique will work for you. It is important to assess the success or failure of an ecotechnological solution on a particular site, before deciding on an appropriate solution, for example, Stangl (2007) investigated the performance of 60 year old hedge brush layers and live crib walls in torrent catchment areas in Italy and Austria. Stangl's (2007) results showed that with increasing age, tree species diversity had decreased, yet there was no loss in soil reinforcement and both methods were found to have excellent soil armouring and anchoring effects.

Traditionally, ground bio-engineering techniques have not been preferentially selected for use in large infrastructure projects, but this does not need to be the case as they can be widely used and applied in large construction projects, as experienced by the Egnatia Odos AE (EOAE) company. The EOAE was set up specifically to manage the design, construction, maintenance, and use of the Egnatia Motorway in Southern Europe. One of their aims was to ensure the environmental protection and land restoration of disturbed land due to construction works. The success of the ecotechnological solutions employed were quantified by Katridzidakis et al. (2007a,b) and Koukoura et al. (2007), and the methodologies employed by Egnatia Odos AE of sourcing locally produced seeds and growing them in their own nursery added to the project's success (Katridzidakis et al. 2007a).

With regard to eco-engineering techniques, careful thought must be given to the instability problem on the slope, how that problem will change over time, and whether the species selected for planting on the site in question will themselves be subject to temporal changes. In the case of e.g. rockfall or windthrow problems on slopes, the forester must also take into account planting density and thinning practices and how the management of a stand with regard to such spatial factors might affect slope mass movement. The combination of ground bio- and eco-engineering techniques should also allow for greater slope stability to be achieved with minimal cost. Nevertheless, it is recommended that professional advice be sought before carrying out any of the applications described in this chapter, as detailed information on installation methods and planting guidelines are not given.

6. REFERENCES

Berger F, Quetel C, Dorren LKA (2002) Forest: a natural protection mean against rockfalls, but with which efficiency? In: Proceedings of the International Congress. Interpravent 2002 in the Pacific Rim – Matsumoto/Japan. Congress Publication, vol. 2, pp 815-826

Coppin NJ, Richards IJ (1990) Use of vegetation in civil engineering. CIRIA, Butterworths, London

Cucchi V, Meredieu C, Stokes A, Berthier S, Bert D, Najar M, Denis A, Lastennet R (2004) Root anchorage of inner and edge trees in stands of Maritime pine (*Pinus pinaster* Ait.) growing in different podzolic soil conditions. Trees – Struct Func 18:460-466

Daniel TW, Helms JA, Baker FS (1979) Principles of Silviculture, Second Edition. McGraw-Hill, New York

Dorren LKA, Berger F, Le Hir C, Mermin E, Tardif P (2005) Mechanisms, effects and management implications of rockfall in forests. Forest Ecol Manag 215:183-195

Dorren LKA, Berger F (2006) Stem breakage of trees and energy dissipation during rockfall impacts. Tree Physiol 26:63-71

Frehner M, Wasser B, Schwitter R (2005) Nachhaltigkeit und Erfolgskontrolle im Schutzwald. Wegleitung für Pflegemassnahmen in Wäldern mit Schutzfunktion. Bern, Bundesamt für Umwelt, Wald und Landschaft: pp 654

Gardiner BA, Stacey GR (1996) Designing forest edges to improve wind stability. Forestry Commission Technical Paper 16. Forestry Commission, Edinburgh

Gardiner BA, Marshall B, Achim A, Belcher R, Wood C (2005) The stability of different silvicultural systems: a wind-tunnel investigation. Forestry 78:471-484

Gray DH, Sotir RB (1996) Biotechnical and Soil Bioengineering Slope Stabilization: A Practical Guide for Erosion Control. Wiley & Sons, Inc., New York

Greenway DR 1987 Vegetation and slope stability. In: Anderson MG, Richards KS (eds) Slope Stability. Wiley, Chichester, pp 187–230

Gsteiger P (1993) Steinschlagschutzwald. Ein Beitrag zur Abgrenzung, Beurteilung und Bewirtschaftung. Schweizerische Zeitschrift für Forstwesen 144:115-132

Hewlett HWM, Boorman LA, Bramley ME (1987) Design of reinforced grass waterways. Report 116, CIRIA, London

Hibberd BJ (1991) Forestry Practice. Forestry Commission Handbook 6. HMSO, London

Hunt R, Gardiner BA (2002) Tree topping. A review of the feasibility of using tree topping to reduce wind damage risk in commercial forest plantations. Internal Report. Forest Research, Roslin

Katridzidakis M, Pipinis E, Kekis G, Ververidou E, Sevastou E (2007) Erosion control by application of hydroseeding methods along the Egnatia Motorway (Greece). In: Stokes A, Spanos I, Norris JE, Cammeraat E, (eds) Eco- and Ground Bio-Engineering: The Use of Vegetation to Improve Slope Stability, Springer, pp 393-400

Katridzidakis M, Pipinis E, Liapis A, Stathakopoulos I, Kekis G, Ververidou E, Sevastou E (2007) Restoration of slopes disturbed by a motorway company: Egnatia Odos, Greece. In: Stokes A, Spanos I, Norris JE, Cammeraat E, (eds) Eco- and Ground Bio-Engineering: The Use of Vegetation to Improve Slope Stability, Springer, pp 401-409

Koukoura Z, Kyriazopoulos A, Karmiros I (2007) Herbaceous plant cover establishment on highway road sides. In: Stokes A, Spanos I, Norris JE, Cammeraat E, (eds) Eco- and Ground Bio-Engineering: The Use of Vegetation to Improve Slope Stability, Springer, pp 387-391

Lewis L (2000) Soil Bioengineering. An Alternative for Roadside Management. USDA Forest Service, San Dimas

Mössmer EM, Ammer U, Knoke T (1994) Technisch-biologische Verfahren zur Schutzwaldsanierung in den oberbayrischen Kalkalpen. Forstl. Forsch. ber. München 145: 135.

Motta R, Haudemand J-C (2000) Protective forests and silvicultural stability. An example of planning in the Aosta valley. Mt Res Dev 20:74-81

Nicoll BC, Ray D (1996) Adaptive growth of tree root systems in response to wind action and site conditions. Tree Physiol 16:899-904

Nyland RD (2002) Silviculture: Concepts and Applications, Second Edition. The McGraw-Hill Companies, Inc. New York

Quine CP, Coutts MP, Gardiner BA, Pyatt DG (1995) Forests and Wind: Management to Minimise Damage. Bulletin 114. HMSO, London

Quine CP, Gardiner BA (1992) Incorporating the threat of windthrow into forest design plans. Forestry Commission Research Information Note 220. Forest Research, Farnham

Ray D, Nicoll BC (1998) The effect of soil water-table depth on root-plate development and stability of Sitka spruce. Forestry 71:169-182

Redfield E (2000) Soil Bioengineering and Biotechnical Stabilization, Renewable Resources 575: Advanced Revegetation, University of Alberta, Canada

Schiechtl HM (1980) Bioengineering for Land Reclamation and Conservation. University of Alberta Press, Edmonton, Alberta, Canada

Schiechtl HM, Stern R (1996) Ground Bioengineering Techniques for Slope Protection and Erosion Control. Blackwell Science Ltd, London

Smith DM, Larson BC, Kelty MJ, Ashton PMS (1997) The Practice of Silviculture, Applied Forest Ecology, 9th edition. John Wiley & Sons, New York

Spanos I, Daskalakou E, Thanos C (2000) Postfire natural regeneration of *Pinus brutia* forests in Thassos island, Greece. Acta Oecol 21:13-20

Stangl R (2007) Hedge brush layers and live crib walls – stand development and benefits. In: Stokes A, Spanos I, Norris JE, Cammeraat E, (eds) Eco- and Ground Bio-Engineering: The Use of Vegetation to Improve Slope Stability, Springer, pp 287-296

Stokes A, Mickovski SB, Thomas BR (2004) Eco-engineering for the long-term protection of unstable slopes in Europe: developing management strategies for use in legislation. In: Lacerda W, Ehrlich W, Fontoura M, Sayao SAB, (eds) IX International Society of Landslides conference, 2004, Rio de Janeiro, Brazil. Landslides: evaluation and stabilisation, AA Balkema Publishers, vol. 2, pp 1685-1690

Wasser B, Frehner M (1996) Wegleitung Minimale Pflegemassnahmen für Wälder mit Schutzfunktion. Vollzug Umwelt, Flankierende Massnahmen (FLAM) des Walderhebungsprogramms (WEP) 1992-1995, Modul Minimalpflege/Erfolgskontrolle. Bundesamt für Umwelt, Wald und Landschaft (BUWAL), Bern

Chapter 8

ECOTECHNOLOGICAL SOLUTIONS FOR SLOPE STABILITY: PERSPECTIVES FOR FUTURE RESEARCH

Alexia Stokes

INRA, AMAP, A A-51/PS2, Boulevard de la Lironde, 34398 Montpellier cedex 5, France

Abstract: *This chapter concludes the work presented in this book. Future research topics are proposed which include suggestions for developing a database of plant species useful for erosion control and slope stability. Decision support systems also need more work and should be available for the general practitioner to use without specialist training. Field studies and numerical modelling should be combined to answer specific questions, especially concerning large-scale slope instability problems. As economics is an important factor to consider, more research into vegetating hotspots only should be performed. Finally, in today's changing climate, carbon sequestration through vegetation strategies could also be an important line of research.*

Key words: decision support systems, databases, modelling, climate change

1. INTRODUCTION

In this book we have tried to review the processes leading to mass movement on slopes and to present the possible mechanisms for mitigating slope failures. A wide range of topics has been covered in the hope that some areas which were not clear to either engineers, ecologists or geographers, have now been made more accessible, and that the relevant literature has been cited and can be accessed further. Nevertheless, some subjects, particularly in the domain of large-scale eco-engineering practices,

J.E. Norris et al. (eds.), Slope stability and erosion control: Ecotechnological solutions, 277–282.

remain poorly understood. For example, the biomechanics of trees subjected to avalanches (Johnson 1987; Ciolli et al. 1998) has received little attention compared to the uprooting of trees during wind storms (see Chapter 4). More research on the management strategies to carry out in protection forests growing in zones at risk from avalanches is a priority and both field experiments and numerical modelling could be used.

The studies presented in this book have largely been focused on knowledge and experience gained through research and applications carried out in Europe. It is absolutely necessary to broaden our research to include countries which are suffering from massive erosion (e.g. China) and repeated landslides (e.g. many tropical countries in south-east Asia and Central – South America). Studies carried out in other parts of the world will also be beneficial for alimenting our own database of information, especially if new ground bio-engineering techniques are contrived which can be adapted for use in other countries.

1.1 Databases

It is absolutely essential that ecotechnologists always use plant species' Latin names. An example of where using the common name can be disastrous was illustrated by Stokes et al. (2007) studying bamboo. The generic "bamboo" has often been cited as being useful in controlling erosion and slope stability (e.g. Storey 2002). However, certain species e.g. *Phyllostachys nidularia* Munro. actually contribute to landslide risk because of their very shallow root systems and tall stems which bend downwards causing buckling and uprooting. Nevertheless, for controlling surficial erosion, this species is useful because of its shallow, fibrous root mats. A database of plant species and how they can be used to fix soil would thus be of major utility in avoiding confusion over species' names. An open-source internet site would also allow practitioners and researchers to update their findings and experience directly. The first step towards such a site can be seen with the Plants for a Future database (http://www.pfaf.org/), which provides habitat information and possible uses of different species e.g. soil reclamation. However, it is not a specialist site, therefore only basic information is provided concerning ecotechnological solutions.

1.2 Experimental data or numerical modelling?

Although numerical modelling has made significant advances in recent years, the information gained from field studies is extremely valuable and can always be used to validate the most complicated models. Nevertheless, modelling enables virtual experiments to be carried out, which would not be possible in the field due to the time taken for vegetation to grow and be stabilized. Pertinent questions which are currently being asked and where modelling can help enormously include: how does vegetation change over space and time and what will be the consequences for erosion and slope stability? The work presented on rockfall protection forests in chapter 7 used a combination of experimental and modelling techniques to successfully answer very similar questions. The planting of forest trees in random, staggered or parallel rows and the influence on soil reinforcement and slope hydrology is the next research area to receive immediate attention (Sakals and Sidle 2004, Kokutse et al. 2006; Danjon et al. 2007). The impact of clear-felling and tree removal in forests has been researched using field data and modelling techniques (see Sidle et al. 2006), but whether to use natural regeneration or to plant the subsequent bare slopes is still not clear. Similarly, more research needs carrying out on the vegetating of defined hotspots only (see Chapter 6). Can these same techniques be transposed to other types of geomorphologically fragile areas? Field studies on plant type urgently need performing, and modelling could be used to determine the effects on localized planting on a large scale (also using e.g. GIS techniques).

1.3 Decision Support Systems

Several decision support systems (DSS) for managing slope stability and soil erosion problems exist already (Dragan et al. 2003; Barac et al. 2004; De la Rosa et al. 2004; Sarangi et al. 2004; Mickovski and Van Beek 2005). GIS modelling of landslide hazard (Lazzari and Salvaneschi 1999), erosion (Huang et al. 2003) and environmental vulnerability (Li et al. 2006), has recently provided DSS tools for water resources management and land use. However, such tools are of limited utility if end-users do not have access to GIS models or data and the necessary expertise to run the DSS (Walker et al. 1995). Therefore, expert and rule-based DSS (Shaffer and Brodahl 1998) lend themselves well for use by end-users who either do not have significant computer expertise; whose needs for information are not continuous or who have little time to spend on elegant but complicated DSS (Crist et al. 2000). Hence, a simple, open-source, rule-based expert system, freely available on the internet, has been developed recently (Jouneau and Stokes 2006). This Slopes Decision Support System (*SDSS*) is based on a DSS previously developed by Mickovski et al (2005) and Mickovski and van Beek (2006). The end user can input data concerning the slope and soil type, vegetation and meteorology of his/her site. The output of the DSS will give a susceptibility index of slope failure, depending on the data and the model used. The *SDSS* is simple and easy to use, even with limited technical knowledge. However, it is a platform, and users must enter their own models, or use those already available in the *SDSS*. For example, FOS could be calculated for a number of vegetated slopes using the software SLIP4EX (see Chapter 5), and simulations could be run to determine the influence of soil, hydrological or vegetation factors. A simple model could then be developed using the results from the simulations (in its simplest form, e.g. a linear regression model) which could then be incorporated into the *SDSS*. Freely available on the internet (http://liama.ia.ac.cn), the *SDSS* is ideal for experts around the world who wish to incorporate their own parameters, models, knowledge and databases. It is hoped that experts and users will adapt the *SDSS* to their own situations, whether that be soil mass movement risk, debris flow, erosion or storm hazard, and then propose to the authors to upload their versions onto the web site, where other users may also access them. The type of user which could benefit from such a system would not just be professionals and engineers, but also local authorities, stakeholders and students.

2. CONCLUSION

To conclude, what we need to ask ourselves now is: how to plant or manage a potentially unstable or degraded site for the *long-term* sustainable conservation of soil? Which species should be used, bearing in mind that in many countries, farmers and stakeholders might need to use this plantation as an income? Where bare land is to be revegetated, should natural regeneration be allowed, or should the soil be planted quickly with young grasses, shrubs or trees? How will vegetation change over time and space? Can forested slopes be clear-felled? Can hotspots only be vegetated or should planting of a degraded zone be more widespread? What are the costs incurred? In today's changing climate, carbon sequestration in soils has become a major issue (Dumanski and Lal 2004): it is necessary to determine whether an increase in plant growth due to elevated carbon dioxide (CO_2) in natural conditions can occur and if so under which circumstances (Körner et al. 2005)? Any such increase may then lead to an increasing input of plant biomass (debris, root biomass and root exudates) in forest and agricultural soils which in turn could cause changes in the carbon budget of soils.

To answer these and more questions, a major international program is needed which covers not only the scientific study of the problem, but also includes the dissemination of results and teaching of new methodology to local decision-makers, authorities and farmers. For example, international research and demonstration sites where techniques to improve degraded lands can be tested, measured, improved and demonstrated to farmers, researchers and community leaders. Decision support systems and databases also need to be freely available to practitioners as well as being easy to use. However, for such a program to be successful in practice, it is vital that local authorities enforce new techniques, but not at the cost of already existing successful systems (e.g. the cutting down of trees to plant new trees). In many countries, more employees should be hired to enforce the environmental acts, regulations developed both now and in the immediate future and offences that damage the ecosystem must be treated seriously.

3. REFERENCES

Barac A, Kellner K, De Klerk N (2004) Land user participation in developing a computerised decision support system for combating desertification. Env Monitor Assess 99:223-231

Ciolli M, Tabarelli S, Zatelli P (1998) 3D spatial data integration for avalanche risk management. In: Fritsch D, Englich M, Sester M (eds) IAPRS, Vol 32/4, ISPRS Commission IV Symposium on GIS – Between Visions and Applications, Stuttgart, Germany pp 121-127

Crist PJ, Kohley TW, Oakleaf J (2000) Assessing land-use impacts on biodiversity using an expert systems tool. Land Ecol 15:47-62

Danjon F, Barker DH, Drexhage M, Stokes A (2007) Using 3D plant root architecture in models of shallow slope stability. Ann Bot-London, in press

De la Rosa D, Mayol F, Diaz-Pereira F, Fernandez E, de la Rosa D Jr (2004) A land evaluation decision support system (MicroLEIS DSS) for agricultural soil protection. Env Model Software 19:929-942

Dragan M, Feoli E, Fernetti M, Zerihun W (2003) Application of a spatial decision support system (SDSS) to reduce soil erosion in northern Ethiopia. Env Model Software 18:861-868

Dumanski J, Lal R (2004) Soil conservation and carbon sequestration (Guest Editorial) Clim Change 65:253-254

Huang YF, Chen X, Huang GH, Chen B, Zeng GM, Li JB, Xia J (2003) GIS-based distributed model for simulating runoff and sediment load in the Malian River Basin. Hydrobiologia 494:127-134

Johnson EA (1987) The relative importance of snow avalanche disturbance and thinning on canopy plant populations. Ecology 68:43-53

Jouneau L, Stokes A (2006) Development of a decision support system for managing unstable terrain: calculating the landslide risk of slopes. In: Marui H, Marutani T, Watanabe N, Kawabe H, Gonda Y, Kimura M, Ochiai H, Ogawa K, Fiebiger G, Heumader J, Rudolf-Miklau F, Kienholz H, Mikos M. (eds) Interpraevent 2006: Disaster Mitigation of Debris Flows, Slope Failures and Landslides. September 25 – 27, 2006, Niigata, Japan. Universal Academy Press, Inc. Tokyo, Japan, ISBN 4-946443-98-3, pp 543-552

Kokutse N, Fourcaud T, Kokou K, Neglo K, Lac P (2006) 3D numerical modelling and analysis of forest structure on hill slopes stability. In: Marui H, Marutani T, Watanabe N, Kawabe H, Gonda Y, Kimura M, Ochiai H, Ogawa K, Fiebiger G, Heumader J, Rudolf-Miklau F, Kienholz H, Mikos M. (eds) Interpraevent 2006: Disaster Mitigation of Debris Flows, Slope Failures and Landslides. September 25-27, 2006, Niigata, Japan. Universal Academy Press, Inc. Tokyo, Japan, ISBN 4-946443-98-3, pp 561-567

Körner CR, Asshoff O, Bignucolo S, Hättenschwiler SG, Keel S, Peláez-Riedl S, Pepin S, Siegwolf RTW, Zotz G (2005) Carbon flux and growth in mature deciduous forest trees exposed to elevated CO_2. Science 309:1360-1362

Lazzari M, Salvaneschi P (1999) Embedding a geeographic information system in a decision support system for landslide hazard monitoring. Nat Hazards 20:185-195

Li AN, Wang AS, Liang SL, Zhou WC (2006) Eco-environmental vulnerability evaluation in mountainous region using remote sensing and GIS – A case study in the upper reaches of Minjiang River, China. Ecol Model 192:175-187

Mickovski SB, Stokes A, van Beek LPH (2005) A decision support tool for windthrow hazard assessment and prevention. For Ecol Manage 216:64-76

Mickovski SB, van Beek LPH (2006) A decision support system for the evaluation of eco-engineering strategies for slope protection. Geotech Geol Eng 24:483-498

Sakals ME, Sidle R (2004) A spatial and temporal model of root cohesion in forest soils. Can J For Res 34:950-958

Sarangi A, Madramootoo CA, Cox C (2004) A decision support system for soil and water conservation measures on agricultural watersheds. Land Degrad Dev 15:49-63

Shaffer MJ, Brodahl MK (1998) Rule-based management for simulation in agricultural decision support systems. Computers Electronics Agric 21:135-152

Sidle RC, Ziegler AD, Negishi JN, Rahim Nik A, Siew R, Turkelboom F (2006) Erosion processes in steep terrain – Truths, myths, and uncertainties related to forest management in southeast Asia. For Ecol Manage 224:199-225

Stokes A, Lucas A, Jouneau L (2007) Plant biomechanical strategies in response to frequent disturbance: uprooting of *Phyillostachys nidularia* (Poaceae) growing on landslide prone slopes in Sichuan, China. Am J Bot 94:1129–1136

Storey PJ (2002) The conservation and improvement of sloping land: a manual of soil and water conservation and soil improvement on sloping land. Volume 1. Science Publishers, Enfield, New Hampshire, USA

Walker DH, Sinclair FL, Kendon G (1995) A knowledge-based systems-approach to agroforestry research and extension. AI Applications 9:61-72

INDEX

FSC
www.fsc.org
MIX
Papier | Fördert
gute Waldnutzung
FSC® C083411

Kolibri 360 GmbH
Ferdinand-Jühlke-Straße 7
99095 Erfurt, Deutschland
produktsicherheit@kolibri360.de